HV	Owen, Tim, 1961- author.
6018	Criminological theory
.O94	30360100039843
2014	

NORTH ARKANSAS COLLEGE LIBRARY
1515 Pioneer Drive
Harrison, AR 72601

Criminological Theory

Criminological Theory
A Genetic–Social Approach

Tim Owen
Senior Lecturer in Criminology, University of Central Lancashire, Preston, UK

HV
6018
.O94
2014

© Tim Owen 2014

All rights reserved. No reproduction, copy or transmission of this publication may be made without written permission.

No portion of this publication may be reproduced, copied or transmitted save with written permission or in accordance with the provisions of the Copyright, Designs and Patents Act 1988, or under the terms of any licence permitting limited copying issued by the Copyright Licensing Agency, Saffron House, 6–10 Kirby Street, London EC1N 8TS.

Any person who does any unauthorized act in relation to this publication may be liable to criminal prosecution and civil claims for damages.

The author has asserted his right to be identified as the author of this work in accordance with the Copyright, Designs and Patents Act 1988.

First published 2014 by
PALGRAVE MACMILLAN

Palgrave Macmillan in the UK is an imprint of Macmillan Publishers Limited, registered in England, company number 785998, of Houndmills, Basingstoke, Hampshire RG21 6XS.

Palgrave Macmillan in the US is a division of St Martin's Press LLC, 175 Fifth Avenue, New York, NY 10010.

Palgrave Macmillan is the global academic imprint of the above companies and has companies and representatives throughout the world.

Palgrave® and Macmillan® are registered trademarks in the United States, the United Kingdom, Europe and other countries.

ISBN 978–0–230–27850–9

This book is printed on paper suitable for recycling and made from fully managed and sustained forest sources. Logging, pulping and manufacturing processes are expected to conform to the environmental regulations of the country of origin.

A catalogue record for this book is available from the British Library.

A catalog record for this book is available from the Library of Congress.

To the memory of my paternal grandparents, Charles William Owen and Gertrude Owen

Contents

Preface viii

Acknowledgements ix

1 Introduction 1
2 Transitions in Criminological and Social Theory 10
3 Constructing a Genetic–Social Framework 63
4 An Application of the Genetic–Social Framework to the Study of Crime and Criminal Behaviour 116

Bibliography 174
Index 200

Preface

The book is intended as a contribution towards metatheoretical development as part of the post-postmodern 'return to' sociological theory associated with Roger Sibeon (1996, 1999, 2004, 2007), Derek Layder (1997, 2004, 2007), Nicos Mouzelis (1991, 1993a, 1995, 2007), Margaret Archer (1995, 1998) and Owen (2006a, 2006b, 2007a, 2007b, 2009a, 2012a, 2012b), in tandem with an attempt to 'build bridges' between criminological theory and insights from evolutionary psychology and behavioural genetics.

The framework employed in the book relies upon methodological generalisations as opposed to substantive ones. It is contented that we require a 'way forward' beyond postmodern relativism, harshly environmentalist oversocialised accounts and genetic fatalism (the equation of genetic predisposition with inevitability), in the form of a modification of Roger Sibeon's ontologically flexible anti-reductionism. It is suggested that this modification would entail focusing upon the biological variable (evidence from behavioural genetics and evolutionary psychology for a, at least in part, genetic basis for some human behaviour). The mutuality between genes and environment is acknowledged in the book, and Matt Ridley's (1999, 2003) model of 'nature via nurture' is favoured.

The explanatory potential of the framework is demonstrated by 'applying' the meta-concepts to the study of the areas of interest. The focus here is upon crime, criminal behaviour and approaches to criminology. It is contended that the metatheoretical framework can usefully contribute to a study of the subject area, and also towards 'preparing the ground' for further theoretical and empirical investigation, drawing upon large-scale synthesis.

Acknowledgements

I would like to record my great debts of gratitude to the following people for their help, support, advice and inspiration in the completion of the book: Professor Jason L. Powell (Manchester Metropolitan University), Professor Keith Faulks (Uclan), Dr Terry Hopton (Uclan), Jessica Marshall (Uclan), Jennah Evans (Uclan), Dr David Scott (Uclan), Dr Roger Sibeon (University of Liverpool), Professor Derek Layder (University of Leicester), Susan Jones, Russell Leavitt, Wayne Noble, Ian Conolly, Mark Owen of SPCS (Preston) for his computer brilliance, Elizabeth Connard and Charlotte Nicklin of Lilibets of Paris (Southport) for their sublime entremets, my sister Wendy Steeple and brother-in-law David, my sister-in-law Louise Ridley, nephew Joseph Ridley and niece Grace Ridley, Harriet Barker and Julia Willan of Palgrave Macmillan and Rajeswari Balasubramanian of Integra. Special thanks are due to my wife Julie.

1
Introduction

This book is a contribution towards metatheoretical development as part of the post-postmodern return to sociological theory and method associated with Archer (1995), Layder (1997, 2007), Mouzelis (1995, 2007), Owen (2009a, 2009b, 2012a, 2012b) and Sibeon (2004, 2007), in tandem with a cautious attempt to build bridges between criminological theory and selected insights from evolutionary psychology and behavioural genetics. In the pages that follow, I suggest a way in which criminological theory might move beyond its four main theoretical obstacles. These obstacles are the nihilistic relativism of the postmodern and poststructuralist cultural turn; the oversocialised gaze and harshly environmentalist conceptions of the person; genetic fatalism or the equation of genetic predisposition with inevitability (Owen, 2009a, 2012a) and bio-phobia (Freese et al., 2003) that appear to dominate mainstream criminology; and the sociological weaknesses of many so-called biosocial explanations of crime and criminal behaviour (see, for instance, Walsh and Beaver, 2009; Walsh and Ellis, 2003), which, although dealing adequately with biological variables, appear to neglect or make insufficient use of meta-concepts such as agency–structure, micro–macro and time–space in their accounts of the person. I suggest that a way forward lies in the form of an ontologically flexible, metatheoretical sensitising device, alternatively referred to as *post-postmodern* or *genetic–social* in order to distance the framework from hard-line sociobiology.

My starting point is to modify Sibeon's (ibid.) original anti-reductionist framework to include a new focus upon the biological variable (the evidence from evolutionary psychology and behavioural genetics for a partial genetic basis for human behaviour in relation to sexuality, language, reactions to stress, etc.), genetic fatalism, the oversocialised gaze and psychobiography. This new framework is capable of making a contribution towards a return to sociologically based theory and method and suggesting a way forward for criminological theory, and also towards a cautious marriage between the biological and social sciences that is balanced and does adequate justice to the mutuality between genes and environment. Here, the evidence that genes play a role alongside environment in terms of causality in relation to human behaviour is considered (Cosmides and Tooby, 1997; Hamer and Copeland, 1999; Pinker, 1994). My contention is that there is sufficient evidence to warrant the incorporation of a focus upon the biological variable into the new metatheoretical framework, alongside meta-concepts, notions of dualism – as opposed to a Giddensian duality of structure – and notions of psychobiography (Layder, 1997, 1998a; Owen, 2009a) that describe the asocial and dispositional aspects of the person, and modified notions of Foucauldian power. The latter notion of modified Foucauldian power entails a recognition of the dialectical relationship between agentic and systemic forms of power; the relational, contingent and emergent dimensions of power; and the concept that *contra* Foucault, power can be stored in *roles*, such as those played by police officers, and in *systems*, the most obvious of which, for criminologists, is the criminal justice system. It is important to keep in mind here the idea of mutuality when focusing upon biological variables in criminological analysis, what we might call the 'feedback loop' which embraces genes and environment, acknowledging the mutuality and plasticity between them. The framework I attempt to develop posits that 'nurture' depends upon genes, and genes require 'nurture'. Genes not only predetermine the broad structure of the brain of *Homo sapiens* but also absorb formative experiences, react to social cues or, as Hamer and Copeland (1999) suggest, can be switched on by free-willed behaviour and environmental stimuli. For example, stress can be caused by the outside world, by impending events, by bereavements and so on. Short-term stressors cause an immediate increase in the production of norepinephrine and epinephrine, hormones responsible for increasing the heartbeat and preparing the human body for 'fight or flight' in emergency situations. Stressors that have a longer duration may activate a pathway that results in a slower but more persistent increase in cortisol. Cortisol can suppress the working of the immune system. Thus, those who have shown symptoms of stress are more likely to catch infections, because an effect of cortisol is to reduce the activity and number of white blood cells or lymphocytes (Becker et al., 1992). As Martin (1997) shows, cortisol does this by switching on genes, and it only switches on genes in cells that possess cortisol receptors, which have in turn been switched on by environmental stimuli, such as stress caused by bereavement. Cortisol is secreted in the first instance because a series of genes, such as CYP17, get switched on in the adrenal cortex to produce the enzymes necessary for making cortisol. There are important implications here which inform my attempt to construct *genetic–social* criminological theory. For example, Filley et al. (2001) have linked elevated levels of norepinephrine with aggressive criminal behaviour. Hostile behaviour can be induced in humans by increasing plasma levels of norepinephrine, whereas agents that block norepinephrine receptor cells can reduce violent behaviour (ibid.). The enzyme monoamine oxidase is involved in the reduction of norepinephrine, and low levels of monoamine oxidase allow norepinephrine levels to increase (Klinteberg, 1996).

My approach to criminological theorising acknowledges that crime may be socially constructed, in the sense that human actors ascribe meaning to the world, but that there is still a reality 'out there', in the sense that environmental conditions are potential triggers of genetic or physiological predispositions towards behaviour that may be labelled criminal. However, that does not mean that behaviour should be viewed as reflecting an inherited, pre-written script that is beyond individual control. For example, reflexive agents possess the agency to choose not to engage in criminal activities where they believe that their actions will harm others and offend ethico-social codes, or where the rewards are outweighed by negative consequences. Agency, in turn, is influenced not only by morality or reason but also by inherited, constitutional variables. An inherited impulsive disposition may predispose an actor to formulate and act upon potentially criminal decisions. In *genetic–social* theorising, the biological variable must be considered as one element within multifactorial explanations for crime and criminal behaviour, alongside a critique of agency–structure, micro–macro, time–space and so on.

This *genetic–social* (Owen, 2009a, 2012a) framework arises in response to what I consider to be the following illegitimate forms of theoretical reasoning: reification, essentialism, duality of structure, relativism, genetic fatalism and the oversocialised gaze. The framework offers a flexible ontology and relies upon a multi-factorial analysis. It is capable of identifying a way forward beyond the anti-foundational relativism of postmodernism and Foucauldian poststructuralism, aspects of our intellectual life that are complicit in the stagnation of critical criminology. An approach which sidesteps the 'nature versus nurture' divide which still haunts mainstream criminology and emphasises instead the mutuality between genes and environment is essential if we are to advance upon Shilling's (1993) starting point for a biological sociology and supersede the biologically top-heavy, largely American attempts at biosocial analysis (Herrnstein and Murray, 1994; Mednick et al., 1987; Mednick and Volavka, 1980; Walsh and Beaver, 2009; Walsh and Ellis, 2003; Wilson and Herrnstein, 1985), which appear to lack a sufficiently sophisticated appreciation of sociological theory that would make them truly 'biosocial'. These elements combined make a framework that can contribute towards a new direction for criminological theory as part of a return to sociological theory and method in the age of the human genome. Its methodological generalisations, as opposed to substantive generalisations, its lack of 'bio-phobia' and its realist social ontology make it a sensitising device with the potential for future theoretical and explanatory use best expressed in terms of large-scale synthesis. My ontological position is, to some extent, influenced by Alain Robbe-Grillet's (1963: 24) Heideggerian repudiation of 'words of a visceral, analogical, or incantatory character', which, one reflects, can arguably be found in Gramscian accounts of hegemony utilised in criminological theorising.

The reader will note that the book consists of four integrated and interlocking chapters. This Introduction is the first of the chapters in which the aims and objectives have been outlined. We now move on to briefly examine the contents of each following chapter and how they fit together. It will be recalled that the ultimate intention here is to develop a metatheoretical framework which will contribute towards post-postmodern theoretical development and towards the building of bridges between the biological and social sciences, and which will also prove its worth in terms of explanatory potential in relation to selected examples of crime and criminal behaviour. We need to examine the individual stages of this process now, as they unfold throughout the following chapters.

Chapter 2 ('Transitions in Criminological and Social Theory') reveals how there has been a mounting reaction in contemporary criminology and social theory against the nihilism and paralysis of anti-foundational, relativistic postmodern and poststructuralist approaches. Sibeon's (1996, 1999, 2004, 2007) anti-reductionist framework is suggested as an initial 'way forward' beyond such relativism, but it is observed that it will be necessary to modify the metatheoretical framework so that it includes new meta-concepts such as the oversocialised gaze, genetic fatalism and the biological variable in order to conceptualise the plasticity of the relations between genes and environment in theoretical analysis pertaining to crime and criminal behaviour, and in order to contribute towards the building of bridges between the biological and social sciences. This modification is essential because Sibeon's anti-reductionism neglects biological variables. The chapter demonstrates how the genetic–social framework (like Sibeon's original framework) is an example of metatheory. Metatheory is designed to equip us with a 'general sense of the kinds of things that exist in the social world, and with ways of thinking about the question of how we might "know" that world' (Sibeon, 2004: 13). It therefore entails a flexible ontology, avoiding the relativism of the cultural turn, and consists of methodological generalisations as opposed to substantive generalisations. Here, I define and cover the 'cardinal sins' of reductionism, reification, essentialism and functional teleology out of which Sibeon's original, anti-reductionist framework, which focuses upon agency–structure, micro–macro and time–space, arose. I examine the merits of a metatheoretical approach which lies in sharp contrast to that of the anti-foundational relativism found in the work of those such as Lyotard (1986–7) and Milovanovic (2013). It is decided that the deficits of postmodern approaches outweigh the merits, and it is reiterated that we require a post-postmodern approach to criminological theorising of the sort employed by Owen (2009a, 2012a). Here, notions of Foucauldian power, often drawn upon by critical criminologists, are also examined. The genetic–social framework employs a modified notion of Foucauldian power, and a demonstration of its usefulness is provided early on in the work, although the 'real' application of a modified concept of Foucauldian power to the area of crime and criminal behaviour is discussed in greater detail in Chapter 4. It is made clear that a modification of Foucauldian power which entails regarding systemic and agentic powers as autonomous and acknowledging the dialectical relationship between systemic and relational powers can contribute towards conceptualising power relations pertaining to crime and criminal behaviour. The chapter then considers issues pertaining to agency and social action, and the importance of the terms agency–structure and micro–macro is established in relation to how they refer to differing dimensions of social reality. As in previous work of Owen (2009a), a section on the work of Margaret Archer, Derek Layder and Nicos Mouzelis follows in which their considerable scholarly contributions towards post-postmodern social theory are discussed, and I consider how the authors conceptualise agency–structure and micro–macro and deal with the question of whether to employ dualism or duality of structure. It is decided that Layder's (1997) concepts of social domains are more closely related to Sibeon's original, anti-reductionist framework. Having examined the concepts and the illegitimate forms of theoretical reasoning (the 'cardinal sins' of reductionism, reification, essentialism and functional teleology) which Sibeon's original framework focuses upon and arises as a critique out of respectively; the problems of relativism and the cultural turn; notions of a modified Foucauldian power; issues pertaining to agency–structure and micro–macro; and the significant contributions of three major theorists towards the post-postmodern transitions in contemporary social and criminological theory, we move towards Chapter 3 ('Constructing a Genetic–Social Framework'). It is important to keep in mind here the idea that it might be necessary to modify Sibeon's original theoretical framework to include a focus upon new meta-concepts such as the biological variable, the oversocialised gaze and genetic fatalism in order to progress to the stage where we have a viable sensitising device with which to conceptualise the relations between genes and environment in the context of crime and criminal behaviour.

Chapter 3 takes the suggestion to modify Sibeon's framework and considers the evidence from evolutionary psychology, behavioural genetics and biological science for, at least in part, biological causality in relation to selected examples of human behaviour. It is the contention here that it is absolutely necessary to do so in order to properly justify the inclusion of new meta-concepts such as the biological variable, the oversocialised gaze and genetic fatalism. Firstly, the genetic–social framework is briefly codified in order to keep the reader on track, so to speak. I then examine the work of several authors who have attempted to straddle the biology/social science divide. In addition to recent attempts by Owen (2009a, 2012a) to apply the meta-concepts of the genetic–social framework to the study of human biotechnology and crime and criminal behaviour, there have been other attempts by Walsh and Ellis (2003) and Walsh and Beaver (2009) to formulate a biosocial criminology, Quilley and Loyal's (2005) application of Elias's figurational theories to the task of synthesising the social and biological sciences, and Freese et al.'s (2003) work on 'biophobia' which are worthy of mention. The main focus, however, is upon the more well-known attempts by social scientists of the embodied or material corporeal 'school' to build bridges between the social and life sciences. Drawing upon the work of the evolutionary psychologist Matt Ridley (1999, 2003), it is demonstrated how genetic predisposition (Ridley employs the phrase 'determinism' in the same sense as predisposition (1999: 307)) need not equate to inevitability because of the mutuality between genes and environment. It is suggested here that an acknowledgement of Ridley's (ibid.) 'Nature via Nurture' model is a possible way forward beyond Shilling's (1993) starting point for a biological sociology, by acknowledging what material corporeal social scientists will not – 'genes do influence behaviour' but they can be 'switched on' by free-willed, external, environmental stimuli (Ridley, ibid.). I next examine the sources of evidence for biological causality with regard to selected human behaviours, focusing largely upon evolutionary psychology. I consider detailed, in-depth criticisms of evolutionary psychology by those such as David (2002), Rose and Rose (2000) and others, and the contrary evidence in favour of the approach from those such as Curry (2003), Fodor (1983), Pinker (1995, 1999), Tooby and De Vore (1987), Daly and Wilson (1998) and others. It is concluded that we can acknowledge the cogent evidence from evolutionary psychology and related disciplines to the effect that it does seem likely that human beings possess instincts in the sense of unlearned patterns of behaviour, as Hamer and Copeland's (1999) work on the biology of the sex drive appears to confirm. Following this, I examine the need to incorporate dualism in the sense of a dualistic conception of biology and the social, alongside Sibeon's dualistic conception of agency–structure. After deciding that a dualistic conception of biology and the social is an effective way of avoiding what Archer (1995: 96) calls 'central conflation' (the author refers specifically to Giddensian tendencies to collapse the distinction between agency and structure), I move on to the question of incorporating the biological variable into metatheoretical analysis and into the genetic–social framework itself. Here, the evidence for biological, or partly biological, causality in relation to selected, specific human behaviours such as sexuality (Hamer and Copeland, 1999), stress (Ridley, 2003) and language (Enard et al., 2002) is more closely examined. In the face of the convincing evidence for a biological component to the above behaviours, the decision to incorporate the biological variable into the meta-framework alongside Sibeon's ontologically flexible notions of agency–structure, micro–macro and time–space, a modified notion of Foucauldian power and dualism, together with an avoidance of anti-foundational relativism and the 'cardinal sins' of reductionism, reification, essentialism and functional teleology, is made. It is also decided to incorporate the new meta-concepts of the oversocialised gaze and genetic fatalism too. It is noted at the end of Chapter 3 that we have now arrived at the point of codifying the new, expanded genetic–social framework and applying it to the study of crime and criminal behaviour.

Chapter 4 does indeed codify the new framework, and each of the meta-concepts is applied to selected examples from the literature pertaining to crime and criminal behaviour in order to illustrate the framework's explanatory potential. Examples include cyber-crime, theories of Globalisation often drawn upon in security studies, the oversocialised perspectives of Environmental Criminology, the reductionism of Marxist Criminology, essentialism in some forms of Feminist Criminology, the misuse of Foucauldian power and the illicit conceptions of the state favoured within Left idealist approaches, the functional teleology inherent in authoritarian populist accounts of hegemony, the neglect of biology within studies of masculinities and crime and so on.

In what follows, I examine recent transitions in criminological and social theory and identify trends towards a post-postmodern criminological and social theory that is better equipped to conceptualise and explain crime and criminal behaviour in the age of the human genome.

2
Transitions in Criminological and Social Theory

There appears to be a mounting reaction in contemporary theory against the 'cultural turn' and the extreme relativism of postmodern and poststructuralist theory. Recently, Hall and Winlow (2012: 8) have drawn attention to the urgent need to 'abandon criminology's weirdly postmodern, self-referential gaze'. The authors cogently refer to the recent trend in criminology towards rejecting or modifying the orthodoxy that crime and social harm are the products of criminalisation and control systems. Scholars such as Owen (2012a), Reiner (2012), Wieviorka (2012), Wilson (2012), Ferrell (2012) and Yar (2012) are bringing causes and conditions back into play, and into criminological analysis. To an extent, it could be argued that there has been a 'return to' sociological theory and method reflected in the work of Mouzelis (1991, 1993a, 1996, 2007), McLennan (1995), Holmwood (1996), Stones (1996), Sibeon (1996, 1997a, 1997b, 1999, 2001, 2004, 2007), Layder (1984, 1994, 1997, 2007), Archer (1982, 1988, 1995, 1996, 1998, 2000) and Owen (2006a, 2006b, 2007a, 2007b, 2009a, 2012a, 2012b). This so-called return to sociology has been the 'accumulation of relatively separate intellectual moves that are a blend of renewed interest in classical sociology and in perennial explanatory problems, together with theoretical reflection arising from critical engagement with comparatively recent perspectives that range from neo-functionalism to actor-network theory' (Sibeon, 2001: 1).

Whilst theorists such as Gellner (1993) view postmodernity (and, in particular, the Lyotardian version) as practically worthless, here the view is that aspects of it are relevant to the development of sociological theory and method (Lemert, 1993; Sibeon, ibid.) in so far as it rejects 'modern' essentialist and reductionist theories. However, despite some useful concepts pertaining to contingency, one concurs with Holmwood's (ibid.) critique of postmodern theory. If, as Lyotard (1986–7) claims, no coherent theory is adequate, how are we to decide which incoherent postmodern theories *are* adequate? Here a 'way forward' is suggested in the form of an anti-reductionist, post-postmodern sociology. We need a form of sociological 'realism', though not in the sense of reductive and reified material evidence. Whilst acknowledging the structural qualities of social systems, an anti-reductionist sociology would recognise that even if social reality *is* socially constructed, 'there may still be a socially constructed reality "out there" ' (Sibeon, 1997a: 3).

Here, I cautiously suggest a modified version of Sibeon's *anti-reductionist* framework consisting of methodological generalisations as distinct from substantive generalisations for the purpose of studying crime and criminal behaviour. The framework is designed to stimulate the development of non-reductive theories and empirical investigations. Above all, the framework is a *synthesis*. Its methodological reach owes much to several theoretical schools mentioned in the Introduction. Sibeon's original *anti-reductionist sociology* arises out of a critique of four illegitimate forms of theoretical reasoning; these are *reductionism, essentialism, reification* and *functional teleology*. A *reductionist* theory is a theory that attempts to explain social 'reality' in terms of a single, unifying principle such as 'patriarchy' (Hindess, 1986a, 1988). *Essentialism* 'presupposes in a prioristic fashion a necessary "unitariness" or "homogeneity of social phenomena" ' (Sibeon, 1997b: 1). *Reification* 'involves the illegitimate attribution of agency to entities that are not actors or agents' (ibid.). *Functional teleology* refers to 'illicit attempts to explain the causes of social phenomena in terms of their effects' (Betts, 1986: 51). My modification of Sibeon's original framework arises out of a critique of the following, additional illegitimate forms of reasoning: *genetic fatalism* and the *oversocialised gaze*. *Genetic fatalism* refers to a widespread tendency within contemporary social science to equate genetic determinism with inevitability. *The oversocialised gaze* refers to 'Environmentalist', sociological accounts characterised by a strong antipathy towards genetic or even partially genetic explanations.

I have also modified Sibeon's original framework so that it entails focusing upon the *biological variable,* in addition to *agency–structure, micro–macro, time–space* and *power* as core concepts. The *biological variable* refers to the evidence from over 25 years of research into behavioural genetics for a, at least in part, biological basis for some human behaviour. This emphasis upon biological variables reflects the influence of insights from recent work which attempts to cross the biology–society divide associated with Newton (2003), Williams (1998, 2003), Bury (1997), Benton (1991, 1994) and insights from the evolutionary perspectives of Daly and Wilson (1998), Tooby and De Vore (1987), Curry (2003) and others. Here, it might be useful to describe a number of the core concepts that originally formed part of Sibeon's *anti-reductionist sociology* and are now included within the 'new framework'. As stated in the Introduction, the 'new framework' has been used to contribute towards a study of human biotechnology (Owen, 2006b, 2009), a post-Foucauldian sociology of ageing (Owen, 2006a), an anti-reductionist approach to ageing (Powell and Owen, 2005) and a study of 'trust' and professional power (Owen and Powell, 2006), and also in the context of approaches to crime (Owen, 2007b, 2012a, 2012b). Here the emphasis is upon a more detailed examination of (selected examples) crime, criminal behaviour and approaches to criminology. In what follows it is essential to examine Sibeon's original framework in detail.

Four 'Cardinal Sins'

Here it is necessary to discuss and shed light upon four 'cardinal sins' – illegitimate forms of theoretical reasoning that are singled out for particular criticism in Sibeon's original framework. Indeed, the original framework arose out of a critique of the terms. *Reductionism, essentialism, reification* and *functional teleology* are also included in my modified framework, alongside additional forms of illegitimate reasoning such as *genetic fatalism* and the *oversocialiSed gaze* (Owen, 2006a, 2006b).

According to Sibeon (2004: 2), whether or not the term 'crisis' has been overused in relation to sociological theory and method, it can be argued that many of criminology and sociology's problems of theory and 'explanatory failure' have been associated with a tendency to draw upon the four forms of illegitimate reasoning: *reductionism, essentialism, reification* and *functional teleology*. Nevertheless, there are reasons for optimism. Despite undoubted difficulties presented by the legacy of postmodernism in terms of *relativism*, there is 'a major drive towards a post-postmodern "return to" sociological theory and method (McLennan, 1995)' which 'seems both feasible and desirable (Archer, 1995, 2000; Mouzelis, 1995; Stones, 1996; Ritzer, 2000; Turner and Rojek, 2001)' (Sibeon, ibid.). What follows is an attempt to define the original four 'cardinal sins'.

Reductionism

Sibeon (ibid.) suggests that *reductionist* theories are ones which attempt to reduce 'the complexities of social life to a single, unifying principle of explanation or analytical prime mover (Hindess, 1986a, 1988) such as "the interests of capitalism", "patriarchy" "rational choice", "the risk society", "trust", "the information society", "globalisation", or whatever'. The author also cites examples in accounts of government and the policy process, theories of the state and so on (theories such as pluralism, elitism or Marxism) which are reductionist, in the sense that each of them is 'predicted on the view that government and public policy can be reduced to a single substantive principle of explanation' (ibid.). Reductionist, 'general' theories are 'ontologically inflexible', resting upon 'a priori assumptions about the nature of the state and of society in regard to factors affecting the distribution of power' (ibid.), and in respect of the nature of political/policy dynamics. Sibeon cogently suggests that 'a more adequate and empirically sensitive form of enquiry' would recognise that certain policy sectors such as education, health and foreign policy, for example, may be 'pluralist', whereas others are 'dominated by policy networks that have an elitist or corporatist form'; and it is possible that 'power distributions and policy dynamics within each sector may shift over time', or they may vary spatially (Sibeon, ibid.). He goes on to suggest, using the present example, that a better understanding of the 'complexities of politics, power and public policy' is likely to be achieved by use of 'non-reductionist, ontologically flexible methods of investigation' such as 'policy network analysis (Rhodes, 1997; Marsh, 1998; Marsh and Smith, 2000) which is a contemporary political science approach' (Sibeon, ibid.: 3). There are, according to the author, three general aspects of his original *anti-reductionism*. Firstly, 'to reject a priori theoretical commitment to analytical prime movers is not to say there are never situations where a very small number of factors' may have causal primacy (Sibeon, ibid.). This should be treated as an 'empirical question' rather than as something, in advance of empirical investigation, that is theoretically predetermined 'on the basis of some reductionist social theory'. It is not suggested that substantive theories should, like 'grand theories', attempt to encompass all the possible factors that may shape the phenomena under investigation. Every substantive theory is partial in terms of the type/number of phenomena to be explained. Secondly, Sibeon distinguishes between 'non-reductive multi-factorial explanation', in other words, explanations which draw upon a cluster of variables and 'compounded reductionism' (ibid.). The latter involves attempts to synthesise/combine two or more reductionisms. The author cites Dominelli's (1997) attempts to synthesise reductionist theories of capitalism and patriarchy, arguing that this results in contradiction and 'explanatory failure'. Thirdly, Sibeon cautions against 'deferred reductionism'. This 'takes an Althusserian-like form' where reductionist explanation is postponed or deferred in social analysis, until the last instance (ibid.). An example, the author suggests, lies in Farganis's feminist critique of postmodern theory, in which a straightforward or obvious reductionism is replaced by a seemingly non-reductionist and multi-factorial approach, which turns out to have a reductionist theory at its base. Another example offered by Sibeon is Harvey's attempts to blend Marxism with postmodernism despite references to contingency. Harvey associates the animation of the social with a 'prime mover', namely modes of production (1989: 107), (Sibeon, ibid.). The approach favoured here also places the *genetic–social* framework at odds with conceptions favoured within Leftist idealist criminology, such as the narrowing of concern to the crimogenic properties of capitalist society.

Essentialism

Essentialism is a 'form of theorising that in a prioristic fashion presupposes a unity or homogeneity of social phenomena', such as the law or other social institutions, or taxonomic collectivities such as 'women', 'men', 'the middle class' or 'white people' (Sibeon, ibid.: 4). The author acknowledges the inescapability of a certain degree of idealisation of phenomena in order to be able to refer to the phenomena at all. However, we should not 'falsely essentialise them or theoretically ascribe to them more homogeneity than they actually possess' (Sibeon, ibid.). Essentialist reasoning 'does not regard the degree of homogeneity or heterogeneity of social phenomena as an empirical variable for investigation' but rather 'presupposes on theoretical or political grounds a "necessary" unitariness of the phenomena' under investigation (ibid.). The author cites Albrow's (1996: 91–2, 94–5) criticisms of essentialist notions of globalisation as highly pertinent. Sibeon's view is that 'essentialist thinking' is linked to or is 'a corollary of a reductionist theory', and like reductionism, essentialism can take on a 'disguised' or 'last instance' form as, for example, when 'women' or 'the working class' are described as social categories that, 'though perhaps acknowledged to be in some respects internally divided and cross-related to other categories', should be regarded as 'ultimately possessing a "primary, overriding commonality" that transcends all other memberships (categories) or affiliations' (ibid.). The author argues that if unitariness is a feature of any social phenomenon, it is to be regarded as a contingent and emergent – and perhaps also a temporary – outcome of social processes rather than a necessary effect of social totality. Foucault's overly holistic concepts, often drawn upon by critical criminologists, arguably have strongly essentialist overtones, which fail to address and acknowledge the discontinuities between macro-cognitive and micro-behavioural spheres of the social world.

Reification

Reification is 'the illicit attribution of agency to entities that are not actors or agents' (ibid.). Drawing upon Harre's (1981) concept of agency and Hindess's (1988: 45) 'minimal concept of an actor', Sibeon argues that an *actor* or agent 'is an entity that in principle has the means of formulating, taking, and acting upon decisions' (ibid.). This is a non-reified definition of a social agent or actor. On this basis, there are only two types of actors: individual human actors and 'social actors' (Hindess, 1986a: 115) or 'supra-individuals', as Harre (1981: 141) terms them. According to Sibeon, the latter are organisations (such as government departments, professional associations), committees, such as the Cabinet, or micro-groups, such as households. The author offers examples of *non-actors*, these being 'entities that cannot express agency', in other words entities that cannot formulate or act upon decisions – such as 'society', 'the state' and taxonomic collectivities such as social classes, 'the British people', 'young people' and so on (ibid.: 5). This non-reified conception of agency challenges the view that, for example, men are an entity (an actor) that can take collective action to remedy gender 'issues'. In so far as agency is a key variable in the production/reproduction of structure, it is incontrovertible that only actors as defined in the original framework can have causal responsibility for social conditions (including inequality), and only actors are capable of formulating and carrying out actions that reproduce or alter those conditions. As Sibeon makes clear, reification, like the other 'cardinal sins' identified in his original framework, is not confined to particular theories/paradigms. Rather, it occurs almost everywhere across the theoretical landscape. He provides a few examples to illustrate this point.

Sibeon cites the work of Touraine on a sociology of action. Sibeon disputes Delanty's (1999: 122–44) description of Touraine as a theorist committed to the 'return of agency' to social theory. He feels that Delanty's overall assessment of Touraine is 'rather too generous', as the latter's 'conception of agency and social action is marred by reification' (ibid.). For example, Sibeon shows how, in Touraine's *The Voice and the Eye*, 'society' is regarded as an actor (1981: 31, 59) and social movements too are regarded as 'actors', as are social classes. In recent work, Touraine's (1995) conception of agency and social action continues to engage in reification, according to Sibeon (ibid.).

In a similar way, Habermas's central theories exhibit tendencies towards reification. For example, Habermas's (1987: 159–60) attribution of agency to social systems is singled out here, specifically the idea that social systems regulate their exchanges with their social and natural environments by means of co-ordinated interventions into the social world. Sibeon also identifies strong tendencies towards reification in the following: Luhman's (1982: 265) attribution of agency to what he has termed 'autopoietic social systems'; John Law's (1991b: 173–4) blending of poststructuralism with actor network theories (formerly translation theories), which leads the latter to the view that 'an agent is a structured set of relations'; Foucault's claims to the ends that discourses are agents (Danaher et al., 2000: 33), and the 'posthuman agency' of Callon (1986), Latour (1988) and Pickering's (2001) use of poststructuralism and actor network theories (Sibeon, ibid.: 6). Pickering and other actor network theorists suggest that the 'natural' material world and objects such as stones can be said to display agency. Jones (1996: 296) rejects such a conception of agency as an 'obscure, hollow metaphysics'. As has hopefully been made clear, the *genetic–social* framework favours conceptualisations of the state as a non-actor, and again this places it at odds with much of Leftist idealist critical criminology and abolitionism. To reiterate, reification is regarded here as the illicit attribution of agency to entities that are not actors or agents. An actor is always an entity that in principle has the cognitive means of formulating, taking and acting upon decisions. Therefore, 'the state' of Marxist analysis, the 'Metropolitan Police' of the McPherson Report, 'society' and so on are not regarded as actors.

Functional teleology

According to Sibeon (2004: 6), this is 'an invalid form of analysis involving attempts to explain the causes of social phenomena in terms of their effects', where 'effects' refer to outcomes or consequences viewed as performances of 'functions'. The author, requesting that we bear in mind his definition of non-reified actor/agent here, argues that in the absence of intentional planning by actors somewhere/sometime, it is a 'teleological fallacy' to attempt to explain the causes of phenomena in terms of their effects (Betts, 1986: 51) (Sibeon, ibid.). Reductionist theorists may begin with what Sibeon calls a 'current social or cultural item' (a social practice, or a law, or a public policy, etc.) and then endeavour to 'work backwards', claiming without any evidence of intentional planning by actors that the item came into being because it 'accorded with the interests' of a taxonomic collectivity such as 'white people' or 'men' (ibid.: 7). Related to a neglect of agency, teleology is flawed by a problem of logic in so far as the factors that bring a social or cultural item into being must predate the existence of the item already existing. Sibeon cites Durkheim in some of his theoretical work as 'guilty' of engaging in functional teleology, although he shows awareness in his methodological work of 'the importance of separating causal explanation from functionalist explanation' (Durkheim, 1982: 90, 95) (Sibeon, ibid.: 7). Famously, Durkheim argued for the need to distinguish between the *causes* of something and the *functions* of something. In Sibeon's view, illegitimate functionalist teleology conflates causal and functionalist explanation and is an attempt to explain the cause of phenomena in terms of the item's 'functions', that is to say, in terms of the item's consequences/effects. There appears to be an absence of evidence for intentional planning by actors 'some place/some time' in the theories of hegemony favoured within criminological theorising by those such as Hall et al. (1978) in relation to authoritarian populism and the idea that Thatcherism had secured hegemony and a new expression of collective common sense, and Connell (2000) in relation to the concept of hegemonic masculinities.

One of the major themes of this book is that Sibeon's original framework, subject to modifications which are outlined in greater detail in the following chapter, can be applied to the study of crime, criminal behaviour and approaches to criminology. The intention in what follows is to demonstrate that the framework is a form of what Sibeon calls 'metatheory', and to discuss in some depth issues such as agency and social actors. We will be drawing upon the work of leading sociological theorists such as Margaret Archer, Nicos Mouzelis and Derek Layder, who are part of the transition in social theory towards a post-postmodern 'return to' sociology.

Metatheory and problems of relativism

Here, the task is to outline the role that metatheory plays within criminological and social theory in general. The rationale for employing metatheory is discussed, and there is a review of controversies surrounding the problem of relativism and postmodern/poststructuralist rejection of theoretical foundations. The framework developed in this book is very much an example of 'sensitising' metatheory, and a development of Sibeon's original *anti-reductionism*.

As Sibeon (2004: 13) suggests, metatheorists are largely 'concerned with ontological questions', including ones such as 'What is "society"? What type of things exist in the social world?' If actors and agents exist, what sort of things are they? 'Does it make sense to employ a stratified social ontology that refers to micro- and macro-spheres or "levels" of society, or is micro-macro – as Foucault, Elias and Giddens claim – a spurious and misleading distinction?' (ibid.). The author also poses the question, 'And when we turn to epistemology, it is important to ask: how is reliable knowledge acquired?' (ibid.). Also questioned is whether 'lay' actors' self-experience is a 'better guide' to 'social reality' than the 'objective' (in theory) perspective of social scientists. Perhaps, the fundamental question, as Sibeon suggests, is whether objective, social reality exists 'out there', independent of our conception. Often the term *sensitising theory* is used in social theory rather than *metatheory*. According to Sibeon, metatheory can and should inform the development of substantive theory, but the types are distinct. Metatheory, such as the framework employed in this book, is designed to equip us with 'a general sense of the kinds of things that exist in the social world, and with ways of thinking about the question of how we might "know" that world' (ibid.). In contrast, substantive theories 'aim to generate new empirical information about the social world' (ibid.). Mouzelis (1993a: 684) makes a distinction between substantive generalisations and methodological generalisations. He notes, as Sibeon (ibid.) points out, that 'their aim is less to tell us things we do not know about the social than to provide us with conceptual tools for asking interesting questions and preparing the ground for the empirical investigation of the social world'. Therefore, the role of any metatheoretical concept (such as the framework employed in this book) is to generate at the meta-level conceptual tools that inform the development of concepts, substantive theories and explanatory schemes and that underpin the design of empirical studies.

As Sibeon (ibid.: 14) notes, 'not all researchers and theorists are happy with the idea of metatheory'. Postmodernists such as Lyotard (1986–7) tend to reject metatheory on the grounds that it is a 'grand narrative'. Owen (2006b) suggests that Lyotard (celebrated in some circles as the postmodern theorist *par excellence*) is an example of a theorist of the 'cultural turn'. Owen raises the question of whether Lyotard rejects foundationalism from which theory can be generated, and whether he provides an acceptable postmodern epistemology and a viable postmodern theory. The most basic criticism of Lyotard's (1973, 1984, 1986–7) relativistic position is that he never applies it to himself, to his own theories and conceptual frameworks. Lyotard is open, that is to say, to *the self-referential objection* (Blackledge and Hunt, 1993) which posits that if all theories are the product of a particular situation, then so too is *that* theory, and it therefore has no universal validity. To put this another way, if truth and falsity do not exist in an absolute sense, then Lyotard's (1986–7) thesis about the relativity of all knowledge cannot be 'true' in this sense. His thesis, however, invalidates itself; Lyotard is arguably hoist on his own petard. In arguing the way he does, Lyotard is surely employing the very criteria of truth and validity which he claims are culturally relative. He informs his reader that norms of truth change and that ideas of a 'good explanation' are matters of 'social convention' (Lyotard, 1986–7: 80). Yet he himself uses such conventions in saying this. He, in a sense, employs reason to try to prove the inadequacy of reason, claiming to provide a universally valid and 'true' explanation of why there is no such thing as a universally valid and 'true' explanation. Lyotard's endeavour is therefore highly problematic and contradictory. As Gurnah and Scott (1992: 148) observe, postmodern theorising rests upon 'a grand narrative.... It thus reproduces precisely those totalising characteristics of which at the level of intellectual critique it is most critical'. Put simply, the postmodern statement that there can be no general theory 'is itself a general theory!' (Sibeon, 1996: 14). There are those who regard postmodernism as a 'dead' theory (Ritzer and Ryan, 2007). However, it may be argued that postmodernism 'lives on' in the criminological theories of Milovanovic (1996, 1997,1999) and poststructuralism in those of Smart (1995) and Young (1995).

Sibeon (2004: 14) makes the point that 'some meta- or sensitising theories formulate "large" generalisations pertaining to common social processes that may be found in a wide variety of social settings'. He gives the example of Giddens's (1989: 295) structuration theory, which is a reasonably well-known form of metatheory, consisting of postulates that are 'intended to apply over the whole range of human social activity, in any and every context of action'. Such *meta*theoretical generalisations of this type are not the same thing as 'grand' generalisations 'associated with reductionist substantive theories such as Marxism, rational choice theory, and radical feminism' (Sibeon, ibid.), theories which attract the criticism of postmodernists such as Nicholson and Seidman (1995: 7). In contrast, Giddens's structuration theory, according to Sibeon (ibid.), is an example of metatheory 'that does not invoke the reductionism associated with "grand" substantive theories' such as radical feminism (which reduces the complexity of social relations to the idea of patriarchal causality); 'structuration theory is ontologically flexible (Cohen, 1987: 279–80, 285, 289, 291), a term that refers to metatheory of a kind' which makes possible the development of a range of competing substantive theories, and for 'open-ended' empirical investigation of phenomena.

As we have observed with Lyotard, postmodernists deny the validity of sociological (or any form of disciplinary discourse) discourse. So, if one wishes to employ a metatheoretical framework such as the one employed in this study, it is essential to address the emergence of relativistic postmodern theory as 'a body of thought that challenges the very notion of social science' (Sibeon, 2004: 15). Reference here is largely towards the poststructural (Baudrillard, 1983) and the postmodern (Lyotard, 1986-7). The discussion is not concerned with versions such as those of Jameson (1991) and Harvey (1989) which draw upon Marxism, or others, such as Lyon (1994) and Kumar (1995) that argue 'that we live in a postmodern type of society' (Sibeon, ibid.). Additionally, the approach favoured here rejects the idea from Milovanovic (2013: 331) that we can delineate between early 'nihilistic', anti-foundational forms of postmodernism and later, more 'mature' forms of 'affirmative postmodernism' which can inform criminological and sociological theorising and offer 'visions for a possible better society'. Arguably, Milovanovic's 'mature' postmodernism rests upon the very same relativism which characterises the 'early' forms such as those of Lyotard. Ultimately, relativism is anti-foundational as is hopefully made clear in what follows. The focus here is upon the aspects of relativistic, postmodern theory which relate to issues of metatheory, agency–structure, micro–macro and so on (central concerns of the study), with the ultimate concern being the development of a metatheoretical framework to contribute towards a post-postmodern criminology. Milovanovic (2013: 330) makes a cogent point in relation to the role of 'chance and indeterminacy' in social interaction, and it is interesting that postmodern criminologists draw upon Quantum and Chaos theories in this respect. Recent developments within criminological theory, such as Constitutive Criminology, which argue for the 'interconnectedness of phenomena and co-production in the manifestation of phenomena', producing 'alternative notions of "cause", "harms", "reduction" and "repression"', and Postmodern Feminism in criminology, which offers 'an "ethical feminism"as opposed to a hate politics, "contingent universalities" and new understandings of the subject in law' together with textual constructions of female offenders and concepts for social justice (Milovanovic, ibid.: 331) are also interesting, but again, they appear to revolve around relativistic theorising and neglect biology in theoretical analysis.

Sibeon (2004) argues that, despite the claims of Gellner (1993), who sees little of value in postmodern thinking, certain aspects are relevant to the development of theory (Lemert, 1993) and to methodological concerns relating to how empirical studies are designed (Fox, 1991). The author regards poststructuralist and postmodern views of contingency (Bauman, 1992; Lyon, 1994: 4), locale and power to have a part to play in the (re)construction of post-postmodern social theory. It was acknowledged above that Milovanovic's postmodern criminology contains a useful acknowledgement of chance and indeterminacy in social interaction. Indeed, there are similarities in Bauman's (1992: 192–6) emphasis upon multiple loci of power, contingency, time–space variability, cultural pluralism and so on as characteristics of postmodernity, and the anti-reductionist, 'flexible – but – realist social ontology' (Sibeon, ibid.: 17) of Roger Sibeon's original framework, which provides the basis of the modified framework employed here for the study of crime and criminal behaviour. Sibeon (ibid.) is in favour of 'putatively non-reductionist postmodern notions of social indeterminacy and time–space variability', provided they are modified along the lines of 'realist' social ontology. They are seen as useful and relevant not only to sociology and criminology but also to political science (Marsh and Stoker, 1995; Hay, 2002) and to policy analysis (Fox and Miller, 1995; Rhodes, 1997). Sibeon suggests that postmodern theory has certain merits in the direction of attempts to criticise 'modern', essentialist theories and the knowledge claims of meta-narratives, and indeed theories of any kind which involve reductionism, reification, functional teleology and essentialism. In general, however, he is of the view that postmodern thought cannot provide a cogent and viable theoretical or methodological basis for social analysis. Again, we can observe in the criminological theorising of Milovanovic (2013) a rejection of the modernist notion of linear development in socio-historical change, for example, Marxian dialectical materialism, Weberian rationalisation and the Durkheimian division of labour.

It is possible to identify the main deficits within postmodern theory, without wishing to embark upon an extended critique, in so far as they relate to the theories of the modified framework employed here, such as essentialism, reification and so on. Firstly, when postmodern critics criticise 'modern' social theories, they tend to equate the latter with the problems associated with the four 'cardinal sins' of reification, reductionism, essentialism and functional teleology. However, one of Sibeon's tasks has been to develop forms of reasoning that avoid the four general 'sins'. The modified framework employed in later chapters to examine selected examples of crime and criminal behaviour develops Sibeon's framework further to regard *genetic fatalism*, which is 'a pervasive tendency within contemporary social science to equate genetic predisposition with inevitability' (Powell and Owen, 2005: 28), and *the oversocialised gaze*, which refers to 'the "environmentalist" accounts characterised by a strong antipathy towards genetics or even partially genetic explanatory frameworks' (Powell and Owen, ibid.) also as 'cardinal sins' of illegitimate theoretical reasoning. Secondly, as Sibeon (ibid.: 18) suggests, there is an 'epistemological contradiction' between the postmodern as a *type of society* and the postmodern as a *type of theory*. In the first example, Sibeon suggests that postmodern writers are asserting that there is something 'real' in the former mode (empirically), exhibiting 'definite structural characteristics (consumerism, extreme cultural pluralism and diversity, social fragmentation, individualism, informationalism, and so on)' (ibid.), and in the latter mode, they claim that 'society is never something that is real but is, rather, an effect of discourse'. So, we are being told, simultaneously, that on one hand postmodern society 'exists', and on the other that it does not. Although postmodernists welcome contradiction and incoherence, there is a problem for those who seek to develop *metatheory* here. It revolves around attempts to combine realist ontology with idealist epistemology. Thirdly, as Best and Kellner (1991) suggest, postmodern theory tends to underestimate the significance of interpersonal relations. Thus, it could be argued that postmodern theory neglects subjectivity and also intersubjectivity. As Sibeon (ibid.) suggests, postmodern theorists have a tendency to lack an adequate theory of agency and of the actor or agent and subscribe to a flawed conception of agency–structure. This is of some importance, as the author argues that interaction and intersubjectivity mediate the relation of agency to structure. Sibeon goes on to argue that postmodern theorists tend to employ deterministic conceptions of the social actor or agent. He sees these conceptions as 'curiously reminiscent' of the determinism (social) and reification of such reductionist, 'modern' theories as Marxism, radical feminism and Parsonian structural functionalism (ibid.). The postmodern claim that knowledge and theories 'embody a standpoint' that has been 'determined by the structural location' of the theorist in terms of his/her membership of social categories is, according to Sibeon (ibid.), a 'determinist or reductionist' conception of agency. In this flawed conception of agency, actors are seen as products or effects of discourses, determined by social contexts. Postmodern theory reproduces the 'cardinal sins' of reification, reductionism, essentialism and functional teleology, and entails an 'objectionist conception of agency and structure' that does not fit neatly alongside an 'idealist epistemology' (ibid.). Fourthly, Sibeon cites relativism and the postmodern view of the relation of discourses to social contents as contradictory. Here Sibeon highlights *relativism* as a major deficit in postmodern theory and 'elevates' it to the 'status' of a 'cardinal sin'. In the modified framework employed here, *relativism* is most definitely codified as a 'cardinal sin'. Seidman (1994: 325) represents a tendency within postmodern theory to claim that 'society or social contexts determine actors' discourses'. This claim is very similar to theories of structural predetermination which posit that actors' cognitive processes are largely determined by their structural locations and by specific social-historical circumstances. Alternatively, postmodernists claim that 'discourses do not refer to any ontologically prior or pre-discursive social reality', as no 'reality' exists under postmodern conditions; instead discourses 'produce or construct social reality' (Sibeon, ibid.: 19). As the author suggests, there appears to be a contradiction here. Postmodernists are arguing both that discourses *produce* or 'determine' social contexts and that discourses are *determined by* social contexts. Sibeon argues that it is 'appropriate' here to endorse the views of Berger and Luckmann (1971), White (1992: 305) and Law (1986a: 304; 1991a: 18), each of whom regard the 'relation of discourses to social contexts' as 'dialectical', though 'without properly recognising that the relation is often only *loosely* dialectical' (Sibeon, ibid.). It could be argued that there is no automatic, direct link between action and structure.

As suggested earlier, whilst certain aspects of postmodern thought contain some useful analytical precepts, especially in relation to contingency and the 'healthy' aspect of incredulity towards metanarratives (Lyotard, 1986–7), the genre as a whole is regarded as 'seriously flawed' by Sibeon (ibid.) and others, such as Rojek and

Turner (2000). Sibeon suggests, for example (whilst acknowledging that there are differences of emphasis among postmodern critics of sociology), that the 'anti-foundationalist' form of postmodern theory (Lyotard, 1986–7) is 'itself a metanarrative' (ibid.). Such relativistic, anti-foundationalism is considered by Turner (1990: 248) to be 'unduly deterministic', failing to analyse the 'epistemological incommensurability between the postmodern as periodisation and as theory'.

Rojek and Turner (2000) outline a critique of 'decorative sociology' as a trend in contemporary social science where 'culture' has eclipsed 'the social'. This is in line with Sibeon's (2004: 22) identification of 'a somewhat whimsical aestheticism that enjoyed its heyday at the height of the postmodern turn' during the late 1980s and early 1990s. Rojek and Turner (ibid.: 1) use the term 'decorative sociology' to refer to 'a branch of modernist aesthetics which is devoted to a politicised, textual reading of society and culture'. The authors seek to 'challenge the political self-image of decorate sociology' as a contribution to political intervention. They acknowledge the contribution of 'the cultural turn' to a revision of the relationships between identity and power, race and class, ideology and representation and so on. However, the authors emphasise that it has done so 'chiefly at an aesthetic level'. Influenced by Davies (1993), they argue that the greatest achievement of the 'cultural turn' has been to teach students how to read material politically. Their hypothesis is that what might be called the 'aestheticisation of life' has not translated fully into the politicisation of culture. Similar to the calls for a return to sociology theory and method (Sibeon, 1996, 1999, 2004, 2007) and Mouzelis (1991, 1995), Rojek and Turner argue that an 'adequate cultural sociology' would have to be 'driven by an empirical research agenda', embrace historical/comparative frameworks and focus upon 'the changing balance of power in Western capitalism' (ibid.). Rojek and Turner maintain that there is a case for advocating the importance of theoretical cumulation as opposed to anti-foundationalism. The authors equate cultural relativism with moral relativism. They suggest (ibid.: 645):

> Because postmodern cultural studies assumes moral relativism, it cannot produce, let alone accept, a unified moral criticism of modern societies. As a result it is intellectually unlikely that cultural

studies would develop an equivalent to Weber's notion of rationalisation or Marx's concept of alienation. Postmodern cultural studies finds it difficult to promote a political version of the modern world apart from an implicit injunction to enjoy diversity and resist commercialisation of popular culture.

It is necessary here to examine postmodern theorists' portrayal of their targets. Postmodernists have tended to depict sociology as dogmatic and over-confident in their rejection of sociological reliance upon metatheoretical, theoretical and methodological foundations. Gurnah and Scott (1992: 41), for example, argue that 'the whole of post-war European and American sociology' exhibits a 'degree of superficiality and smug sense of infallibility'. In response, Sibeon (2004: 22) argues that sociology in its modern form is actually characterised by 'theoretical pluralism and by epistemological uncertainty and reflexivity', and that these characteristics have accelerated since the 1950s. The work of Gibbs (1989) also supports this claim. According to Swingewood (2000), the history of sociological thought is characterised as much by intellectual caution and introspection as it is by smugness and certitude (Calhoun et al., 2002). As Sibeon points out, Seidman's (1992) argument along the lines that to formulate sociological perspectives based upon explicitly articulated theoretical or metatheoretical premises ('foundations') is the same as suggesting that all 'competing sociological perspectives' should be denied the right to exist (Sibeon, ibid.). Seidman goes on to argue that 'foundational rationales are never more than local, ethnocentric prejudices' (1992: 60). As Sibeon (ibid.) points out, Seidman then proceeds to develop a 'foundational rationale' for his own views.

It is arguably possible to avoid the four 'cardinal' sins referred to by Sibeon, without having to engage in postmodern hypercontextualism. In later chapters we see the development of a flexible, metatheoretical framework employed to examine aspects of crime and criminal behaviour, and the desire here is to follow Sibeon in avoiding the crude either/or choices between ontologically rigid and dogmatic 'modern' theories, and the equally flawed relativistic approaches of postmodern and poststructuralist theories. The approach advocated here follows Sibeon's work and entails a pliable ontology which includes a processual conception of 'the social' as potentially variable across time and social space. The ontology, though flexible, is 'realist'; that is to say, it favours a conception of a heterogeneous, shifting empirical reality 'out there', which rejects the hypercontextualism and relativism of postmodern/poststructuralist epistemology. As referred to earlier, Lyotard's postmodern repudiation of 'metanarratives' is, in part, a rejection of reductionist, single-order 'modernist' theories which emphasise, for example, capitalism or globalisation as universal principles of explanation. It may be possible to agree with Lyotard's criticisms of 'modern', essentialist theories and to borrow his sense of 'conceptual flexibility', whilst at the same time insisting upon approaches based upon ontological and epistemological 'realism', which are rejected by postmodernists. In Sibeon's (ibid.: 23) view, therefore, 'a pliant but realist ontology' would recognise that we can conceptualise foundations in terms of (meta)theoretical postulates pertaining to social processes and mechanisms to do with agency, structure, time–space and social chance rather than to empirical structures, events or social patterns. Sibeon considers Giddensian structuration theory to be 'ontologically flexible' (ibid.), referring to Cohen's work (1987: 279–80, 285, 289). Cohen (1987: 297) observes that, in regard to Giddens's assumptions about macro-social processes, 'consistent with the ontological flexibility of structuration theory at large, Giddens holds open for substantive enquiry all questions regarding specific systemic patterns' as well as the degree to which social systems are 'stable, organised and permeable'. The type of ontological 'relation' which Cohen refers to here, as does Stones (1996), is not the same thing as what Sibeon (ibid.: 24) calls 'crude modernist theories' which replicate the 'cardinal sins' of reification, reductionism, essentialism and functional teleology.

Postmodern criticisms of foundations in sociological theory relate to questions raised concerning the relation of social discourse to social contexts. Criminological and sociological theories are arguably disciplines and subjects in their own right, and as such may be modified and improved upon. As Sibeon (ibid.: 4) makes clear, this does not have to conflict with the 'parallel view that sociology does not stand in grand isolation' from 'lay' discourses (Friedrichs, 1972: 298). It can be argued that the relation of sociological discourse to 'lay' discourse and to social conditions is 'loosely dialectical' (ibid.), described by Giddens (1987: 32) as 'the double hermeneutic'. There is a suggestion here, by Sibeon, that the concepts/theories of sociology 'spiral' in and out of social life. As the author suggests, when sociologists investigate the 'empirical world', actors' meanings and concepts 'will tend to wind their way' into sociological discourse. Sibeon (ibid.) cites juvenile crime as an example, explaining how studies of the sociological area draw upon/refer to concepts of criminality, self-responsibility and so on that are employed by 'relevant actors' such as young offenders, police officers and so on. Actors' meanings, therefore, 'permeate, and are crucial resources for, social scientific accounts' of the empirical world (ibid.). Sometimes, sociological concepts are 'diffused' to settings outside of academe. Mouzelis (1995: 52) has noted, according to Sibeon (ibid.), the extent to which particular social science discourses have/do not have impacts upon social contexts, as 'an empirical variable', though this is something postmodernists and poststructuralists might question in view of their belief that the social world described by academics is entirely a product of discourse. Sibeon goes on to say that, whilst there may be no 'direct link' between academic and 'lay' discourses, there 'tends to be a *loosely dialectical* linkage' to the extent that each kind of discourse influences the other to some degree, which is 'a matter for empirical assessment' in each case (ibid.: 25). Here, we must note that Sibeon refers to a theory which forms part of the discussion of agency and social actors later in this chapter: the idea that the relation of social science discourse to social contexts is such that each is partly shaped by the other but neither fully determines the other – each has (relative) autonomy. An acknowledgement of the idea that social science discourse 'spirals' out and into the 'social fabric' is *not* to suggest that social scientists cannot formulate general propositions about social life (White, 1992: 5, 304). Rather, Sibeon argues that we should reject the relativism of Gurnah and Scott (1992) and Seidman (1992, 1994), reflecting that

> the extent to which such propositions embody (or do not, as the case may be) any identifiable cultural, regional, ethnic, political or gender-related nuances or specificities of time and place, and, where any of these are shown to exist, their implications for social science knowledge, are matters that in each instance merit collaborative academic assessment and reflexive response. Taken together, these considerations suggest that archetypal absolutist and positivist scientism, or the hypercontextualism and relativism

of poststructualism and postmodernism, are not the only epistemological alternatives: relatively impersonal generalised categories of social science knowledge are possible.

(Alexander, 1992: 323; Sibeon, ibid.: 25)

In relation to the controversies surrounding the concept of theoretical 'foundations', a very important proposition (in epistemological and ontological terms) is that metatheory, substantive theories and empirical data should be compatible with each other, and should influence each other. Indeed, it could be argued that 'meta' or sensitising theory, such as Sibeon's original anti-reductionism, or the modified version applied here to study crime, should be seen as *revisable*, 'at the point where substantive theory and empirical data intersect', and 'the relation of this intersection to metatheory should be regarded as a dialectical relation' (ibid.). In theory, then, 'theory-data reciprocity' should stand in a 'mutually regulating relation to metatheory', and theory, data and metatheory should 'shape each other' (ibid.). In certain respects, Sibeon's original anti-reductionist metatheory, from which the framework employed here develops, is similar to Margaret Archer's (1995). Archer's 'morphogenetic' social theories are discussed in greater depth later in this chapter as relevant to the transitions in criminological and social theory that have led to post-postmodern 'synthetic' approaches. Archer, as Sibeon (ibid.: 26) makes clear, makes a distinction between (her terms) Social Ontology, Explanatory Methodology and Practical Social Theory. Firstly, she suggests that these should be 'mutually regulative' and consistent (Archer, 1995: 28), and that it is an advantage to have a clear social ontology/metatheory as metatheoretical assumptions of the ontological kind are impossible to avoid and influence substantive theory as the 'likelihood of ontological contradictions' (Sibeon, ibid.: 26) arising at the level of substantive theory is reduced by an explicitness at the level of metatheory. As Archer has observed, (which Sibeon (ibid.) concurs with), the possession of an 'explicit' epistemology does not guarantee against all errors. Archer warns against 'conflationary' conceptual schemes; there may be a consistency there between social ontology, methodology and 'practical' social theory yet because they are 'conflationary', such schemes can be of no explanatory use. 'Conflation' refers to the compression of two or more variables, so that their individual idiosyncratic qualities are obscured. Sibeon's original anti-reductionist framework, focusing upon agency–structure, micro–macro and time–space, and arising out of a critique of reductionism, essentialism, reification and functional teleology, has an 'affinity' with Archer's framework, and also with Layder's (1998a, 1998b: 92) theory of social domains. Archer's 'analytical dualism', opposed to the duality of structure and action, is similar to Sibeon's position regarding 'duality of structure', in that it seeks to maintain the distinction between agency and structure. This 'duality of structure' concept is included in the framework developed in later chapters to examine crime and criminal behaviour.

This chapter deals with some of the core concepts and concerns of Sibeon's original framework. Sibeon's original framework incorporates Foucauldian insights pertaining to power. Whilst it could be argued that the main focus is upon agency–structure, micro–macro and time–space, Sibeon (1996, 1999, 2004) certainly draws upon Foucauldian ideas pertaining to power. However, whilst acknowledging that Foucauldian (1972, 1980a, 1982) concepts relating to power, if used selectively, can serve as a useful corrective against exclusively systematic perspectives (Sibeon, 2004: 135), there is a tendency in Foucault's work (1972, 1980a, 1980b, 1982, 1991) towards *relativism*, and in particular a 'failure to incorporate a non-reified concept of agency into social analysis as part of a dual system and integration approach' resulting in illegitimate teleological forms of explanation (Sibeon, 2004: 73). The point of this section of the chapter, concerning metatheory and problems associated with the 'cultural turn', is to highlight problems around *relativism* in particular and to emphasise the need to move beyond postmodern and poststructuralist relativism towards a *post-postmodern* framework. Drawing upon Sibeon (2004), Owen (2006a: 193) suggests that 'Foucault's conception of *power* may be successfully incorporated' in analysis and metatheory, and this must be done in a critical and selective fashion:

> informed by the critique of *agency-structure, micro-macro* and other terms offered in the 'new' framework. If we are to employ Foucauldian insights pertaining to *power* we must recognise the *dialectical* relationship between *agentic* and *systemic* forms of power; the *relational, contingent* and *emergent* dimensions of power,

and the concept, that *contra* Foucault, aspects of power can be 'stored' in positions/roles and in social systems/networks.
(Mouzelis, 1995; Owen, ibid.)

It is necessary here to closely examine Foucauldian analysis, both because it employs a *relativism* which the framework here and the frameworks employed in the work of Owen (2006a, 2006b) reject, and because, as stated previously, Foucauldian notions pertaining to *power* may be employed in a modified form to study selected examples of crime and criminal behaviour.

Foucauldian 'Power'

Foucault's 'overly-holistic' concepts (Sibeon, 2004: 71) have strongly *essentialist* overtones that fail to address and acknowledge the discontinuities between macro-cognitive and micro-behavourial spheres of the social world (Sibeon, ibid.). There appears to be a tendency in Foucauldian theory to 'collapse' the distinctions between *agency* and *structure* and between *micro* and *macro*, resulting in what Archer (1995) calls 'central conflation'. Foucauldian insights tend to rest upon a *'reified* concept of *agency'* (Sibeon, ibid.: 74). Foucault's theorising tends to incorporate 'reified concepts of agency and the concept of *duality of structure* into analysis, with the result of creating *teleological* forms of explanation (Owen, ibid.). Armstrong (1987: 1217), drawing upon Foucauldian insights, warns against too readily endorsing alliances between "bio-social perspectives", perhaps displaying the *genetic fatalism* and *over-socialised* approach of such Foucauldian analysis' (Owen, ibid.). Foucault's (1972) suggestion that sexuality is purely a socio-cultural creation is explored in greater detail in Chapter 3, with contrary evidence provided by the work of Hamer and Copeland (1999). However, despite some of the theoretical deficits outlined, *power* is one aspect of Foucauldian analysis, which

> if used in a 'modified' and reflective way, can certainly contribute towards the study of crime and criminal behaviour. *Power* in the Foucauldian sense, is, 'the milieu in which individuals and groups operate'; the workings of power are not centred in any one group

or source, and do not arise from any given location in the social structure or operate from any singular site.

(Owen, ibid.)

Power is e*verywhere* – the network of surveillance, patterns of discipline and knowledge that serve them have emerged outside of any social actors' control. These ideas can be used as a corrective against exclusively systemic perspectives.

One concurs with Sibeon (2004: 135) to the ends that the Foucault-influenced 'Actor-Network' theories of Callon and Latour (1981) contain persuasive arguments in favour of power having 'no single or prime cause, but that strategic success in the acquisition of power is always potentially reversible'. For Sibeon (ibid.), we should borrow elements of Foucault's ideas pertaining to power in a 'critical, selective fashion' informed by critique of *agency–structure, micro–macro* and *time–space*. It is useful to 'recognise the relational and emergent aspects of power, but power also has *systemic* qualities' (Owen, ibid.: 23–4). Latour, influenced by Foucauldian insights, claims that 'power is not something you may possess or hoard', and that power is 'an effect...never a cause' (1986: 265). Sibeon (ibid.: 136) cogently disagrees with Latour's suggestions, arguing that 'Latour is wrong; power can be hoarded or stored, and therefore power – though often an effect – can sometimes be a "cause"'. One tends to agree with Sibeon that Foucault and 'Actor-Network' theorists such as Callon and Latour (1981) and Latour (1986) tend to 'push their relational and processual conception of power to the point of denying that power can be "stored" in roles and in social systems and networks of social relations'. As will be seen in Chapter 3, we may therefore incorporate a synthesis into the metatheoretical framework which combines Foucauldian and other *relational* concepts of power with an understanding that there is also a *systemic* dimension to power. An acknowledgement that 'some social agents "possess" more power than others (Best and Kellner, 1991: 70), and that the reason for this may lie in the fact that certain elements of power can be "stored" in roles, social systems etc, is something which Foucauldian theory tends to ignore' (Owen; ibid.). For an application of these ideas to Garland's 'culture of crime control' thesis, see Owen's work (2012a), as in his view these 'roles' can include those of police officers and magistrates and dominant networks of social systems.

Particularly useful for the sensitising framework developed here is the idea that there may be multiple forms of power, including systemic power (associated with power storage) in discourses, social institutions, social positions/roles, and in social systems and agentic power which refers to a capacity of agents. These *systemic* and *agentic* powers should be conceptualised as autonomous, though they may influence each other and agentic power

> may be of a relatively contingent, emergent form. Such agentic power may interact with systemic power (role/position) that is 'stored' within what Foucault terms 'discourse', and social systems. In other words, we may think in terms of a *dialectical* relationship between *systemic* and *relational* forms of power.
>
> (Owen, ibid.: 910–11)

Hopefully, at this point, the need to move 'forward', beyond the relativism and nihilism of the 'cultural turn', and the deficits of Foucauldian analysis identified here, towards a metatheoretical framework which focuses upon *agency–structure, micro–macro,* aspects of Foucauldian *power* and *time–space* has been established. In Chapter 3, the argument for including a focus upon *biological variables* (evidence from behavioural science and evolutionary psychology which supports, at least in part, biological causality regarding some human behaviour) is explored. In what follows, we examine in greater depth *agency* and *structure* and *micro* and *macro,* together with the links between the theoretical approaches of certain contemporary social theorists (Roger Sibeon, Derek Layder, Nicos Mouzelis and Margaret Archer) and the implications for both post-postmodern criminology and for the development of the framework. Chapter 3 contains a codification of a suggested metatheoretical framework, which is 'applied' to selected aspects of crime and criminal behaviour in the final chapter (Chapter 4).

Agency and social actors

In what follows, the intention is to establish in general terms the importance of the concepts of *agency–structure* and *micro–macro* in terms of how they refer to differing dimensions of social reality, and to review certain conceptual dilemmas and arguments surrounding the contemporary debate of these concepts in social science. On the basis of a critique, suggestions are offered as to how we might employ these concepts in a developed framework. We have already cautiously criticised the relativism of the 'cultural turn' and the Foucauldian concept of *power* previously. It should be noted that we are dealing here with *modern* social and criminological theory, and despite the fact that some of the theories draw upon classical theorists (Marx, Weber, Durkheim, etc.), the intention is to focus upon how *agency–structure* and *micro–macro* are handled by contemporary theorists. Here, we are concerned with major contemporary theorists such as Anthony Giddens, Margaret Archer, Nicos Mouzelis and Derek Layder.

It is arguably not enough simply to say that Lyotard (1986–7) and other postmodern/poststructuralist critics neglect agency if one is concerned with the development of post-postmodern, metatheoretical framework. As Sibeon (1997a: 5) suggests, what is required is a 'non-reductionist and non-reified definition of agency'. In Sibeon's original, *anti-reductionist* terms, agency is defined by means of Hindess's (ibid.) definition (discussed previously) in which actor/agent is 'a locus of decision and action', where the action is in some sense a consequence of the actor's ability to formulate and act upon decisions. However, there are some entities that cannot under any circumstances be or become actors. These include taxonomic collectivities such as 'men', 'women', 'white people', 'black people'; social classes; 'society' and 'the state'; and objects such as money, written materials, natural disasters, diseases and so on. Additionally, none of these categories of phenomena can be or become agents. It is the view here that 'human beings are reflexive agents with the agency to choose not to engage in criminal activities where they believe that the rewards are outweighed by negative outcomes or actions offend moral prohibitions' (Owen, 2012a: 94). Agency, in turn, is influenced by inherited constitutional variables.

In social theory, criminology and in social sciences disciplines in general, *agency* tends, as Sibeon (ibid.) points out, to be associated with 'human creativity' and social chance (contingency), and *structure* with 'patterned relations, with constraints upon behaviour and with macro-social phenomena'. '*Micro*' is 'for the most part employed as a term that denotes settings of face-to-face interaction', whilst '*macro*' is 'frequently used to refer to "society" and social institutions' (ibid.). Often we may find that *agency* is linked to *micro*, with *structure* associated generally with macro-social phenomena. Sibeon's original anti-reductionist framework views some of the conceptualisations as problematic. Nevertheless, we must recognise that they enjoy quite wide support and there is a general (though not universal) tendency among social theorists to accord considerable importance to distinctions between agency and structure, and micro and macro. As Sibeon suggests, it is the view of certain contemporary theorists that agency–structure (Archer, 1988: ix–x) and micro–macro (Knorr-Cetina, 1981: 2; Munch and Smelser, 1987; Ritzer, 1990) are '*the* core underlying problematics of social theory and sociology' (ibid.: 36). The author has no major disagreement with such views, but adds that '*social choice*' (*contingency*) and '*time–space*' are also highly significant concepts. The latter term is incorporated into the framework in Chapter 3, alongside agency–structure and micro–macro. *Contingency* is arguably another term for *social chance*. Sibeon's original *anti-reductionism* portrays social life as relatively fluid, processual and *potentially* indeterminate unless and until segments of social life are spatially/temporally stabilised and institutionalised for short or long durations of time (Hindess, 1986: 120–1). This raises an important question. As Sibeon (1997a: 68) maintains, there are arguably no valid reasons for constructing an 'either/or theoretical dichotomy between, on the one hand, a structural conception of rigid system determination and predictability' and, on the other, a 'conception of the social fabric as a process of endless flux', and random change. Elias's (1978: 162) figurational sociology paints a picture of social reality as indeterminate, continuous and processional, as an unfolding of 'figurations in constant flux with neither beginning nor end'. One could argue that Elias is utilising an illegitimate dichotomy which is neither statis nor flux. Sibeon's original anti-reductionist framework transcends this dichotomy through the empirical application of a theory which recognises that social life is always potentially open to change and variability, the extent to which stability and continuity, or else discontinuity and change, actually occurs being treated as a variable for empirical investigation. Social conditions are viewed as being *contingently* produced and reproduced as an effect of the self-reproducing tendency of social systems and networks, and in part as the consequences of actors' decisions and actions. As Clegg (1989: 17) indicates, because the social fabric is not structurally predetermined it is, at least, potentially indeterminate and variable. In the concept of *'time–space'* a core assumption is that material which travels in time must also travel spatially, and that spatial dissemination necessarily has a temporal dimension. Cohen (1989: 77) analyses Giddens's structuration theory, observing:

> [I]f social patterns are embedded in the reality of social activity, then a concern for time and space becomes difficult to avoid. Social conduct, after all, is always situated in specific settings, and it takes time to engage even in the most fleeting practices, let alone sustained sequences and series of interaction.

Arguably, Cohen's observation highlights the fact that temporality is a very significant variable among competing sociological theories. As Clegg (ibid.: 212) suggests, 'Different theoretical perspectives diverge on the temporality within which the conceptualisation of action is achieved.'

As mentioned previously, there are theoretical approaches in modern social theory (some of which draw upon classical sociology) which tend to emphasise *structure* whilst neglecting *agency*, according analytical primacy to the macro-social. *Structuralism* is one such approach. Here, there is an analytical movement from subject/actor to structure. In other words, people are conceptualised as 'products' of structure, as 'decentred' from their own meanings. As Sibeon indicates, structuralists view structure rather than agency as the 'main focus of analysis' (2004: 39). The theoretical shift from agency to structure is based on the idea that social structure is self-regulating and synchronic. As Scholes (1974: 10) has said, 'At the heart of the idea of structuralism is the idea of system: a complete, self-regulating entity that adapts to new conditions by transforming its features whilst retaining its systemic structure.' Such a statement entails *reification* – one of the 'cardinal sins' repudiated by Sibeon's original framework, and incorporated into the new framework (codified in Chapter 3). As Sibeon (ibid.: 39) goes on to say, this *reification*, together with 'determinism based on a reductionist notion of language and action', is a feature of a subdivision of structuralism, known as structuralist linguistics (Saussure, 1974; Levi-Strauss, 1963,

1974). Sibeon cites the work of Sharp (1980) as an example of structural linguistics' objectivist, reified conception of language. The 'determinate supposition that language is an autonomous entity' existing independently is part of the structuralist idea (found in Barthes, 1967) that human actors are 'inserted' into language (Sibeon, ibid.). Similarly, in structural linguistics, there is no 'correspondence between a word and the object to which the word refers' (ibid.). Rather, the 'meaning' of a word is determined by its relation to other words, and to concepts of objects, but not to material objects in 'the real world'.

It is not only structuralism and its offshoots which engage in the deterministic neglect of agency. *Poststructuralism* (we have previously discussed Foucault's poststructuralist ideas pertaining to power) is also implicated. The work of Lacan (1977) and Foucault (1970, 1972, 1980a) is offered as examples by Sibeon. Structuralists argue that 'neither signifier nor signified represent or are linked to anything "real", but second, that there is a more or less stable link between signifier and signified' (Sibeon, ibid.: 39). Poststructuralists such as Lacan and Foucault would reject the second proposition. According to poststructuralism, texts and indeed human society are 'open' to a multiplicity of 'readings', and 'any interpretation or reading is never "finished" or "final" but is and should be subject to endless challenges and re-interpretations' as can be seen in the work of Derrida (1982) (ibid.). 'Reality' itself is denied, and analysis must centre upon discourses and the construction of a 'putative reality' (ibid.). Sibeon gives the example of the poststructuralist concept of *intertextuality*, which refers to the 'interplay of texts, which results in texts modifying each other when they are read and re-read (re-interpreted)', and thus texts, and by implication any 'discourse' that refers to the 'real' world, 'are indeterminate and subject to continuous re-interpretation and reformulation', and there is no 'correspondence theory' criterion for arbitrating among interpretations or discourses' as poststructuralists reject the idea that there is a 'real world' against which 'competing discourses or representations can be empirically tested' (ibid.: 39–40). As noted in the sections of this chapter concerning 'Metatheory and Problems of Relativism', postmodern and poststructuralist critics not only engage in *relativism*, but their approach to *agency* is, arguably, deterministic, viewing the social actor as 'decentred', as a 'product' of discourse.

Sibeon (ibid.: 40) gives Laclau and Mouffe's (1985, 1987) work as an example of *Discourse Theory*, which borrows in part from Saussure's structuralist linguistics. This latter form of theory, drawing upon postmodern and poststructuralist insights, posits that discourses and societies are never 'closed' – there are no 'fixed' or stable meanings – but rather, they exist in a state of endless reformulation and flux (Sibeon, ibid.: 40). *Discourse Theory* includes the following poststructuralist ideas: 'there is no pre-discursive social reality and therefore no benchmark against which we can empirically test the veracity of competing discourses or representations of "reality"' and 'any discourses which attempt to convey a sense of the "real world" are arbitrary, and endlessly reformulable' (ibid.). Actors' forms of thought and actions are merely the 'products' of discourse. One tends to agree with Sibeon's criticism of such concepts not only in terms of the neglect of agency and social determinism on the part of *Discourse Theory* but also because of its refusal to take seriously the concept that there exists an ontologically complex and real empirical world that is at least partly independent of discursive activities.

There are other examples of contemporary theorists (deterministic approaches) that tend to either neglect agency or elevate concepts of structure above agency. Sibeon (ibid.: 40–1) describes Habermas's (1986, 1987, 1989) theoretical schemes as another example of an 'over-emphasis on macro-structural phenomena and a neglect of agency'. According to Sibeon, Habermas attempts to 'combine voluntaristic action theory with a deterministic systems theory'. Whilst acknowledging that 'non-contradictory syntheses of action and structure are possible', Sibeon (ibid.) claims that Habermas does not achieve this. Because Habermas's (1987: 159–60) 'systems are imbued with agency', they are thereby *reified* (ibid.). Further, the author draws attention to

> Habermas's contradictory tendency to largely ignore agency in the study of what he calls 'system' (polity and economy). In his theoretical framework, system is inspected only in system-integration terms, whereas he argues, the 'lifeworld' (taken for granted meanings employed in everyday life among family, friends and associates) should be viewed in social-integration terms: this

neglect by Habermas of agency and actor–actor relations in politics, bureaucracies and markets and conversely, the neglect of systemic, role-patterned, and institutional aspects of the lifeworld, limits the usefulness of his approach to agency-structure.

(ibid.: 41)

Viewed in terms of the definitions of *agency–structure* and *micro–macro* used by Sibeon, and incorporated into the 'new' modified framework employed here, Bourdieu (1977, 1984, 1990, 1998), widely regarded in his lifetime as a major theorist of agency–structure, is 'ultimately condemned to a deterministic view of agency' (Sibeon, ibid.). Bourdieu's theory of 'constructivist structuralism' is 'much closer to structuralism than to constructivism', as he believes 'actors' dispositions are largely determined by the social positions they occupy and by their 'habitus' (ibid.).

In what follows, we address some themes, controversies and links between the approaches of a selection of contemporary theorists pertaining to debates around *agency–structure* and *micro–macro*. As Sibeon (2004: 45) has observed, 'a certain amount of ambiguity and also variation in meaning are features of the debate surrounding agency-structure and micro-macro', and 'misinterpretations occur quite often'. One example is Mouzelis's (1991: 35) criticism that Giddensian structuration theory ignores 'macro-action' (which Sibeon (ibid.) describes as meaning 'decisions made by people in positions of authority such as government ministers or heads of large firms' as well as 'decisions made by meetings or committees whose members have a high level of formal authority' such as meetings of heads of state). Sibeon (ibid.) argues that Mouzelis's criticism, repeated in his (1993a: 682) later work, rests on a misunderstanding on the part of the critic himself. According to Sibeon (ibid.), Mouzelis's critique is flawed due to his misunderstanding of 'how the theory of structuration relates to what Mouzelis (but not Giddens) terms the "micro-macro" distinction: nowhere, despite Mouzelis's claim to the contrary, does Giddens deny the existence of what Mouzelis rather confusingly calls "macro action" '. Another example of misinterpretation in contemporary debates around agency and structure lies in what Sibeon (ibid.) refers to as 'McLennan's (1995: 121) erroneous claim – made in a paper that in other respects is scholarly and insightful – that Mouzelis (1991) criticises Giddens on the grounds that Giddens associates micro with voluntarism and with agency, and macro with structure viewed as constraint upon agency'. As Sibeon makes clear, Giddens (1984: 139; 1993: 7) rejects the conflation of micro-agency and macro-structure, and 'it is also the case that Mouzelis (1991: 32) recognises that Giddens opposes these conflations', and here McLennan seems to have misinterpreted the theoretical positions of both Giddens and Mouzelis.

Aside from differing ways of dealing with *agency–structure* and *micro–macro*, and misapprehension of terminologies, there are 'divergences' that rest on 'real differences of approach at the level of social ontology' in the case of a selection of contemporary theorists examined here (Sibeon, 2004: 46). Sibeon identifies the morphogenetic social theory of Margaret Archer as containing an 'inadequate conception of micro-macro' (ibid.). The majority of contemporary theorists appear to use 'micro' to refer to small-scale social phenomena or units of analysis. In contrast, Archer (1995: 8–12) argues that micro–macro 'should not refer to differences in the absolute "size" of social phenomena, but rather, to *relative* differences in size and to a relational conception of scale associated with the concept of "emergence"' Sibeon (ibid.). Archer (ibid.) suggests that a given unit of analysis may be 'micro' in relation to one aspect or stratum of society, and 'macro' in relation to others: 'what justifies the differentiation of strata and thus use of the terms "micro" and "macro" to characterise their relationship is the existence of emergent properties pertaining to the latter but not to the former' (Archer, 1995: 9). Thus, as Sibeon (ibid.) points out, in Archer's terms, a dyad can be viewed as micro, but if it formed part of a 'slightly larger social grouping (a committee or a household, say), then the latter, in relation to the dyad, would be investigated as a "macro" phenomena'. Sibeon recommends resisting Archer's conceptualisation of *micro-macro* (though he has 'no quarrel with the general idea of emergence') in favour of more conventional definitions. A reason for Archer's desire to abandon such conventional distinctions between micro and macro lies in her belief that '*micro-macro*' and '*agency-structure*' are 'simply versions of the same debate' (1995: 7). Sibeon (2004: 47) disagrees, arguing that 'micro is not at all the same thing as agency, and macro is not the same thing as structure'. If we draw upon critique of the four 'cardinal sins' of the author's original framework, employing a flexible social ontology, we clearly see that

> agency is defined in terms of an explicitly anti-reificationist and minimal concept of actor; social structure is defined, in similarly 'minimal' fashion, as 'social conditions' (or the 'conditions-of-action'); and micro-macro refer to the units and scale of analyses concerned with the investigation of varying temporal and spatial extensions of the social. In particular, it was emphasised that social structure is a contingently reproduced set of social conditions, not a necessary effect of the social totality or something that is historically predetermined.
>
> (ibid.: 57)

As Sibeon (ibid.: 47) suggests, another motive behind Archer's unorthodox handling of *micro–macro* is her assumption (1995: 10–11) that those who use the term 'micro' to refer to small-scale interaction and phenomena 'tend to regard the micro-social sphere of interpersonal relations as insulated from the macro-social sphere'. Sibeon cites Knorr-Cetina and Cicourel (1981); Munch and Smelser (1987); and Ritzer (1992) as examples of theorists who employ the micro–macro distinction to indicate differences in properties/scale of phenomena, whilst insisting on 'precisely the idea of links between micro and macro' (ibid.). Most theorists appear to be opposed, therefore, to what Archer (1995: 10) refers to as an '"isolated" micro world'. Whilst acknowledging the contribution Archer has made, especially with regard to her morphogenetic social theory, Sibeon recommends that we reject Archer's version of micro–macro. The *'micro–macro distinction'*, provided it is not employed in the same way as in Archer's work, and in ways which avoid the 'cardinal sins' of *reductionism, reification, essentialism* and *functional teleology*, is

> a useful conceptual tool for marking out variation in the properties and temporal and spatial scale of social phenomena, for denoting corresponding differences in the nature and size of the units of analysis employed in social enquiry, and the exploring links between these units of analysis.
>
> (ibid.: 48)

Archer (1988, 1995, 1998) is very much involved in contemporary debates around 'dualism versus duality', which, as Sibeon (ibid.) indicates, 'centres mainly though not exclusively upon critique of Giddens's theory of structuration (Giddens, 1981, 1982, 1984, 1991b, 1993)'. In what follows, we examine the contribution to this debate, which is in itself related to agency–structure and micro–macro, by selected contemporary theorists. In Chapter 3, the term 'duality of structure' is 'officially' included in the 'new' modified version of Sibeon's original framework. Certainly, Sibeon most definitely favours *dualism* rather than *duality* in relation to *agency–structure* and *micro–macro*. In the following chapter, there is a more in-depth consideration of *dualism* and *duality*, the implications for the framework and an application of the concept of *dualism* to the 'building of bridges' between criminology and biology. It will be remembered, from the introduction and the beginning of the chapter, that the intention here is to develop Sibeon's framework further so that it entails focusing (amongst other things) upon the *biological variable* (evidence for, at least in part, biological causality regarding some human behaviour) alongside *agency–structure, micro–macro* and *time–space*. In Chapter 3, we identify a tendency to argue for a *duality of structure* amongst writers of the 'embodied' school, in which 'biology' and 'society' are seen as 'two sides of the same coin'; for example, Shilling's (1993) analysis of the body as *simultaneously biological and social*. Newton (2003: 35) sees 'no epistemological reason to erect a barrier between the biological and the social'. These positions, in the field of those who seek to build links between the social and life sciences explored in the next chapter, appear to be very similar to Giddens's (1981, 1982, 1984, 1991b, 1993) theory of structuration, in the sense that they are 'elisionist' (Archer, 1995), collapsing 'biology' and 'society' together as Giddensian 'flat' social ontology collapses the micro–macro distinction. Giddens is not the only theorist to effectively collapse this theoretical distinction. As Sibeon (2004: 48) makes clear, so do others such as Elias, Foucault, Callon and Latour, and poststructuralist and postmodern theorists. Giddens has suggested that 'tensions' between micro-theorists and macro-theorists arise out of a 'phony war' (1984: 139) and that micro–macro is a false and unhelpful dualism (1993: 4) 'that polarises social scientists into proponents of two opposing approaches' (Sibeon, ibid.); he also thinks that the distinction helps perpetuate a tendency of some theorists to link micro with agency, and macro with the constraints 'applied' by social structure upon agency (1984: 139; 1993: 7). Another factor in Giddens's unwillingness to employ these terms is that 'micro–macro' distinctions are dualisms serving to emphasise differences between small groups and larger phenomena, such as organisations. Archer's work is similar, but she rejects Giddens's elision and substitutes her 'idiosyncratic version of micro-macro dualism' (Sibeon, ibid.). For Giddens (1979: 204–5), it is the distinction between 'face-to-face interaction in situations of co-presence' (Sibeon, ibid.) and inter-relations with other actors who are spatially/temporally 'absent' that assumes greatest importance. Put briefly:

> Giddens favours the idea of a *duality* – not a *dualism* – of action and structure. In structuration theory the 'duality of structure' is a concept which insists that agency (or action) and structure are not separate domains, but instead are 'two sides of the same coin': the notion of duality specifies that structure is not external to or apart from action – unless structure is currently being practiced (or instantiated) by people, it has no current existence (other than as 'memory traces' in people's minds).
>
> (ibid.: 48–9)

Later in this chapter, we will see how Archer is critical of what she terms 'central conflation'. Archer has Giddens's conflation of agency and structure (see above quotation) in mind (1982, 1988, 1995, 1996). Derek Layder and Nicos Mouzelis, also discussed in greater depth later in the chapter, are also critical of approaches which conflate *agency–structure* and *micro–macro,* though Mouzelis tries to accommodate both duality and dualism in his theoretical approach. Giddensian 'methodological bracketing' emphasises a procedure that rests on a distinction between 'institutional analysis', which refers to the study of macro-historical processes, and 'strategic conduct analysis', which Giddens (1984: 375) defines as 'social analysis which places in suspension institutions as socially reproduced, concentrating upon...actors' (Sibeon, ibid.: 49). As Sibeon shows, Giddens argues that it is very necessary to employ such a distinction and to concentrate analysis on either agency *or* structure. Giddens's argument, in effect, is that even though in reality action and structure are not ontologically separate or different phenomena, it is in practice essential to treat them as distinct, separable phenomena. Several critics, for example Derek Layder (1984: 215; 1998b: 101), have suggested that despite arguing for a *duality* of structure at the ontological level, Giddens effectively 'lets in by the back door' an ontological *dualism* through his analytical strategy of 'methodological bracketing' (separating action and structure). Sibeon comments that

> the dualisms of agency-structure and of micro and macro are indispensable in social analysis; we should, as Layder (1994) aptly puts it, seek to build bridges between agency and structure (and between micro and macro) rather than theoretically collapse the distinctions by compressing the elements so tightly together that they cannot be separated. It follows that as well as my being critical of Giddens's insistence upon a 'duality' of action and structure, I am unconvinced by the rejection of dualism in Law's (1994) 'sociology of ordering' (which is a synthesis of structuralism, post-structuaralism, actor network theory, and, to a much lesser extent, symbolic interactionism).
>
> (Sibeon, ibid.: 49–50)

Sibeon (ibid.: 50) acknowledges Law's focus upon social networks (influenced by Foucauldian insights) and recursion arguably has some merits, but the author disagrees with Law's attempts to dissolve the micro–macro division (Law, 1994: 11, 18, 138), and his attempt to dissolve the agency–structure distinction (Law, 1994: 158–60, 103, 138). According to Sibeon, Law considers the maintenance of the agency–structure divide as 'synonymous' with essentialist/reductionist reasoning. Sibeon (ibid.) considers Law to be 'profoundly mistaken' here (ibid.). Law draws upon structuralist and poststructuralist material in which the 'de-centring' of subject (1994: 24) is advocated, and like Foucault views agents as effects of discourse (ibid.: 113). In Law's theorising there can be no dualism of agency and structure, because Law has wrongly assumed that the agency–structure distinction is redundant by virtue of agency being variable, contingent and relational.

There are further complications which arise from the way certain contemporary theorists deal with the micro–macro division. Nicos

Mouzelis (1995: 123) contends that Anthony Giddens's rejection of the division between micro and macro is based upon a 'belief that social scientists tend to erroneously link micro with agency, and macro with structure'. This is not true of *all* social scientists, and Mouzelis argues that it is therefore unfortunate that Giddens rejects the dualism, and in doing so abandons a useful analytical method. With *dualism*, one may examine the specific characteristics of either variable (agency, structure, micro, macro, etc.) rather than entertaining the possibility of missing salient points pertaining to each individual variable by engaging in an approach which favours *duality of structure* in analysis. Arguably, Giddens has tried to replace micro–macro with his own, personal conception of 'the social integration/system-integration distinction' (Sibeon, ibid.: 50). In Giddensian terms, a social integration type of analysis focuses on face-to-face ('co-presence') relations between actors, and system integration analysis is concerned with 'larger' settings, such as relationships across time and space. As Sibeon (ibid.) points out, Mouzelis (1995: 123) has suggested that this had led Giddens to erroneously view face-to-face interaction/relations as 'minor phenomena', whereas in Mouzelis's (1991: 83,109; 1995: 124) view (the concept of 'macro-action'), such relations between powerful actors should be classed as macro-phenomena as the decisions taken by such actors at, for example, the level of head of state may have 'repercussions that stretch widely across time–space and that may affect the lives of literally millions of people'. There are grounds for agreeing with both Giddens's and Mouzelis's refusal to link micro with agency and macro with structure, and also with Mouzelis's rebuttal of Giddens's treatment of Lockwood's (1964) social/system integration distinction. However, it is possible to challenge Mouzelis's argument to the ends that Giddens erroneously links micro with 'co-presence', suggesting that it is better to remain with the idea that face-to-face interactions are micro-phenomena. Mouzelis appears to view 'co-presence' as a 'lesser' feature, rather than a defining feature of micro-interaction between actors, since interaction in specific cases involving powerful actors, such as the head of state, or of some transnational corporation, for example, often involves decisions with macro-consequences, in the sense that their effects may be felt across time and space. Sibeon's (ibid.) view is that Mouzelis's arguments here are 'unconvincing', preferring to 'define *all* situations of co-presence – irrespective of the power of the participants – as micro-social phenomena', and adding that 'micro-social phenomena should be regarded as relatively autonomous of, though not completely detached from or unaffected by, macro-social phenomena, these being empirical questions'. The intention here, it must be noted, is not to stray too far into the vast literature upon governance, but it is useful to cite Sibeon's (ibid.) example of how, if we were to apply Mouzelis's terms of governance in the European Union, 'his conception of "macro action" (or "macro events") would apply to, for example, interactions in the Council of Ministers or in the European Commission'. Mouzelis's notion of 'macro-action' would also apply to governmental institutions, such as the US Department of Energy, and non-governmental (made up of 2,000 scientists at 20 institutions in six countries) institutions, such as the International Human Genome Sequencing Corporation (HGSC), were we to apply his concept to the 'public' Human Genome Project. However, Mouzelis's formulation contains some serious deficits. Firstly, the 'emergent micro-situational properties of a face-to-face situation' (an ad hoc meeting, for example) may 'significantly influence the course of interaction and thereby shape the decisions taken', and empirical examples can be found in the work of Russett and Starr (1996: 242–3) (Sibeon, ibid.). Although Mouzelis recognises interactional–situational dimensions of social action, he does not relate this to the concepts of 'macro-events' and 'macro-action'. Rather, as Sibeon (ibid.) suggests, Mouzelis's references to powerful actors involved in 'macro-events' involve great emphasis upon 'positional power' (1991: 91; 1995: 24), as if he believes that personal interaction/power dynamics in the interaction and relations of 'top dogs' (such as heads of state, etc.) 'can have no significant emergent micro-processual or relational dimensions that affect decision-making'. Additionally, Mouzelis

> wrongly downplays the extent to which 'macro action' involves far more than an initiating decision made by a few powerful leaders; social action and processes leading up to and during decision-making, and afterwards in the implementation phase, are often embedded in systems of power that entail a very large number of 'routinised circumstances of co-present interaction'.
> (Giddens, 1993: 7; Sibeon, ibid.: 51–2)

In regard to these factors, it could be argued that such a hierarchical concept of macro-action, which does not seem to take into consideration any notion of social chance (contingency), is problematic, and perhaps oversimplified. Thirdly, as Sibeon (ibid.: 52) suggests, the events which Mouzelis (1991: 90–1) describes as 'micro-events', where decisions are taken by less 'powerful', 'micro actors' (1991: 144; 1995: 27, 120), may turn out 'to have far reaching ("macro") outcomes; for example, the seeds of radical political transformation' may, in principle, 'be found in encounters among "ordinary people" ', depending upon the circumstances and unintended consequences, contingency and so on. Fourthly, some meetings of committees with membership of powerful actors may actually be routine, without any meaningful decision-making. Such meetings could not be labelled 'macro-events', in terms of Mouzelis's formulation, energised by 'macro-actors'. Sibeon goes on to argue that Mouzelis's examples of 'macro-events' and references to 'macro-actors' 'rest on a largely systemic, role/positional conception of power' (ibid.). This is in contrast to Foucauldian notions of relational power. Indeed, Sibeon (ibid.) is of the view that Mouzelis (1991: 75, 83, 90–1, 168) displays 'a tendency to adopt, in regard to the matters under discussion here, a mechanical and overly systemic view of power'. Finally, against the formulations of Mouzelis, writers such as Parsons (1995), Rhodes (1997) and Dye (1998), concerned with governance/public policy, have observed that 'big' decisions taken by 'powerful' actors, or 'top dogs', can have only minor consequences. The work of Richardson (1996) and James (1997) contains examples of instances where technical implementation difficulties or successful resistance by implementation agents such as administrators or professionals and contingency can throw public policy off course. As stated previously, the intention here is not to review the literature on governance and public policy, but to note that Sibeon's (ibid.) application of Mouzelis's formulations to the field is worthwhile and these formulations are useful illustrations of the theoretical deficits inherent in such hierarchical concepts of power. Traditional mechanisms of hierarchical, 'top down' government are in some policy sectors declining in significance in the face of non-hierarchical mechanisms of societal steering and co-governance (Kooiman, 2003; Pierre and Peters, 2000).

So far in this chapter, we have examined the four 'cardinal sins' of *reductionism, reification, essentialism* and *functional teleology* out of which Sibeon's original framework arises as a critique. We will note that Chapter 3 deals with ways of developing the 'sensitising device' further by including such concepts as (defined in the Introduction and at the beginning of this chapter) *genetic fatalism*, and the *oversocialised gaze*, as 'cardinal sins', and including a focus upon '*the biological variable*' alongside *agency–structure, micro–macro, time–space* and *power*. We have looked at metatheory (made up of meta-concepts such as agency and structure), and considered the need to adopt ontologically flexible, post-postmodern 'sensitising devices' in the light of the inadequacies of postmodern and poststructuralist theories which entail extreme *relativism*, hyper-contextualism and idealism. We have also noted that aspects of Foucauldian notions of power are conceptually useful if we recognise the dialectical relationship between *agentic* and *systemic* forms of power; the *relational, contingent* and *emergent* dimensions of power, and the concept, that *contra* Foucault, aspects of power can be 'stored' in positions/roles and in social systems/networks (Mouzelis, 1995) (Owen, 2006a: 193). Later, we examined, in some depth, the meta-concepts of *agency–structure* and *micro–macro* in relation to debate around conceptual controversies and ambiguities within contemporary social theory. We looked at structuralism, poststructuralism and discourse theory as selected examples of deterministic approaches which serve to elevate one or another conception of structure above agency. The works of Habermas and Bourdieu were also examined in relation to the same tendency. What Sibeon (2004: 56) has termed 'miscommunication and conceptual inconclusiveness' as well as disagreement in recent, contemporary theoretical debates pertaining to *agency–structure* and *micro–macro* has been examined too, in relation to selected examples, such as Mouzelis and Archer. It is clear that we need to utilise Sibeon's original framework in the light of transitions in social theory, in a contribution towards a post-postmodern 'return to sociological theory and method', and to develop it further (which is attempted in Chapter 3). In what follows, we examine the work of Nicos Mouzelis, Margaret Archer and Derek Layder in greater depth. All three theorists are significant figures in the recent movement towards a renewal of sociological theory and have made incisive inroads into unresolved theoretical problems (such as the relationship of agency to structure) which criminology and sociology need to address in order to retain theoretical/empirical explanatory 'credibility' in the post-postmodern era. Comparisons between theoretical formulations are made, including between the approaches of Layder, Mouzelis and Archer's work and the concepts of Sibeon's original framework which is developed further here, and which is used to study crime and criminal behaviour.

Archer, Layder and Mouzelis: Post-postmodern theorists

Margaret Archer

As Owen (2009a) has suggested, Archer's greatest contribution to contemporary social theory centres upon her (1982, 1988, 1995, 1996) 'morphogenetic social theory'. Archer (1995: 1) has made it clear that 'the "morpho" element is an acknowledgement that society has no pre-set form or preferred state' and the 'genetic' part is a 'recognition that it takes its shape from, and is formed by, agents, originating from the intended and unintended consequences of their activities'. In her (1982) work, the author makes it clear that *morphogenesis* refers to the elaboration of structural forms, and *morphostasis* to their maintenance. Sibeon (2004: 97) has 'reservations' about certain aspects of Archer's framework. As we have seen, Archer's unorthodox notion of *micro–macro* has come under criticism from Sibeon (ibid.) However, the latter author values Archer's anti-conflationist formulations and dualism, adding that Archer 'has much to offer the future development of sociological theory and method'.

Part of Archer's theoretical framework involves the repudiation of what she terms 'downward conflation', in other words, sociological analysis which depends upon methodological collectivism. As Sibeon (ibid.) points out, in Archer's (1995) work pertaining to downward conflationist theorising, *agency* is explained *in terms of* structure. Sibeon gives the example of Talcott Parsons' later work, in which the 'central value system is said (with reference to structure and social system) to induce compatibility among indications and (in terms of agency) to socialise agents so that they energise the social system' (ibid.: 97). According to Archer (1995: 2, 16), downward conflation rests upon a less powerful 'structural conditioning', and in downward conflationist paradigms actors are portrayed as unreflective, socialised beings lacking the kind of inventive capacities that might shape structure. Archer (1998: 83) observes that, in such theories, 'there is little or no sense of society as something which in its existing form may be desired by no-one'. Society, in Archer's formulations, is the 'unplanned outcome of inter-group conflict, negotiation and compromise' (ibid.). Archer identifies as 'downward conflationist' the idea associated with some Marxists, structuralists, functionalists and so on that society 'results from' in-built, systemic forces. Conversely, 'upward conflation' is associated by Archer (ibid.) with interpretative sociology, in which society is viewed as what Sibeon (ibid.: 97–8) terms 'an aggregation of micro (face-to-face) interactions'. Archer (1995: 60) associates 'upward conflation' with interactionist sociology, with 'the neo-phenomenological school' (ibid.: 84), and with methodological individualism (ibid.: 34–46). As Sibeon (ibid.: 98) observes of Archer:

> She argues that upward conflationist ontology wrongly treats structure as the product of agency exercised by current as distinct from previous actors, and wrongly assumes that structure is no more than 'other actors' and their activities. She observes that these postulates erroneously disallow the proposition that structures have aggregate and emergent properties that are 'more than' the sum of the interacting individuals and their decisions and actions, such properties having a shaping or 'conditioning' influence upon actors' forms of thought and activities (and very often, structure operates as a major constraint upon certain types of action); relatedly, upward conflation theorists tend to see structure as readily alterable by actors, providing they have the necessary motivation and information to promote social change. In both upward and downward forms of conflation, relative autonomy is withheld from agency and from structure: primacy is erroneously given to agency *or* to structure as the ultimate constituent of society, rather than investigating 'the two-way interplay between them'.
>
> (Archer, 1998: 74; Sibeon: ibid.)

Archer rejects another form of theoretical reasoning, which she terms 'central conflation' (1995: 167–8), and which she associates above all else with the structuration theory of Anthony Giddens. Archer rejects Giddensian 'duality of structure', and drawing upon the work of Bhaskar (1989a, 1989b), insists that individual social actors and society are different kinds of phenomena, rather than being 'two sides of the same coin'. As Sibeon (ibid.: 98) points out, Archer (1995: 101) considers Giddens's concept of 'social practices' to be flawed, in the sense that it does not 'do justice' to agency or structure, with both viewed as bound together so that there is no possibility of investigating how they interplay, or of ascertaining their respective levels of importance in a particular situation (Archer, 1998: 75, 81). Archer's analytical dualism (in which agency and structure are viewed as distinct, separable variables) suggests that temporality is central (1995: 92) to the mutually shaping relation between agency and structure.

Archer's main criticism of Giddension structuration revolves around the latter's inability to separate agency and structure in a way that would, in a sense, 'open them up' in order for us to analyse them separately, in any social context. For Archer, Giddens is an 'elisionist' (Archer, ibid.: 93) who collapses agency and structure together – structure has no existence outside of its agentic instantiation.

Sibeon (ibid.: 99) makes clear his respect for Archer's 'clarification of the grounds for rejecting upwards, downwards, and central conflation', viewing it as 'invaluable' for social analysis. Also viewed as 'invaluable' is her 'flexible but realist social ontology allied to her epistemological and methodological arguments; her work on a stratified conception of the actor; and her work on "positional interests" and on the dialectics of agency and structure'. 'Positional interests' refer to vested interests. However, one aspect of Archer's framework is open to criticism. Archer's (1995: 147, 148, 253) view of the 'activity dependence' of social structure is based on the argument that 'current structures' are the effects of actions taken by actors who are, in her words, 'long dead'; therefore, there is a 'temporal escape' of social structures from the actions of the past. As Sibeon (ibid.: 100) points out, Archer argues that the activity dependence of, as an example, demographic structure is 'past tense': in her words, 'here the activity dependence of such structures can be affirmed in only one acceptable way: by reference to the activities of the long dead' (Archer, ibid.: 143). In general terms, Archer claims that 'we are all born into a structural and cultural context which, far from being of our making, is the unintended resultant of past interaction among the long dead' (ibid.: 253). Archer's claim that present structures are effects of *past* actions is actually connected, ironically, to a confusing conflation of both the relation of present-day actors to current structures and the relation of present-day actors and structures to previous actors and previous structures. Archer appears to conflate them in a way that incorrectly downplays the extent to which present-day social contexts depend upon the activities of current actors. Arguably, Archer acknowledges that current activities in aggregate form have the effect of unintentionally reproducing society. Yet, Archer (ibid.: 72), in a discussion of the activity dependence of human societies, appears to suggest that current actors play no significant part in proceedings. Sibeon (ibid.) argues that Archer 'is right to refute methodological individualism as a form of upward conflation' (1995: 77; 1998: 76–8), but we must keep in mind the 'extent to which the activity dependence of current contexts or structures may relate to current activities'. Archer (1995: 145) has argued that activity dependence of social structures can be 'present time *or* past tense'. Thus, as Sibeon (ibid.) points out, Archer (ibid.: 145) is perhaps incorrect in arguing for the existence of present-day structures that are 'ontologically *independent* from the activities of those people here present'. As Sibeon (ibid.) makes clear:

> It is legitimate to refute methodological individualism, to argue that structures have emergent properties, and to observe that the activity dependence of present structures relates to past actions; but these statements should not be taken as far as Archer's extreme claim that some current structures may be *entirely independent* of current actors. It seems that here, in a way that sits uneasily alongside some of the implications that may be thought to arise from what Archer (ibid.: 141) elsewhere calls the 'no people; no society' truism, she wants to identify situations where, in her view, there is a total separation of current structure from current action.

Nicos Mouzelis

For Owen (2009a), Mouzelis is certainly one of the leading figures in the contemporary movement towards a return to sociological theory (for example, Mouzelis, 1989, 1991, 1993a, 1993b, 1994, 1995, 1997, 2000). Mouzelis's (1991) work may be regarded as a 'landmark' (Sibeon, ibid.: 101) in the movement to reject the relativistic nihilism of postmodern/poststructualist theory in favour of a 'return to' sociological theory and method. In what follows, we examine an abridged review of selected aspects of his theoretical work. With regard, for example, to Mouzelis's extremely pertinent contributions to the 'duality/dualism' debate (covered briefly in part in this chapter's section on 'Agency and Social Actors'), this is covered in greater detail in Chapter 3. For the time being, we will examine Mouzelis's theoretical work pertaining to the three dimensions of social action.

Mouzelis's theoretical typology contains significant treatment of agency–structure and micro–macro (1991: 196–200; 1995: 136–7), and 'distinguishes between role/positional, dispositional, and situational-interactional dimensions of social action' (Sibeon, ibid.), and the emphasis is upon focusing upon each of these dimensions in social analysis. Mouzelis (1991: 106) uses the example of the 'role scripts' of a teacher and pupil to illustrate his typology, showing how, as Sibeon (ibid.) acknowledges, activity and interaction in the school classroom are 'shaped' by the roles/positions of participants, at least in some cases. Mouzelis's *dispositional* dimensions of social life borrow from Bourdieu's (1977, 1990) theory of 'habitus'. 'Habitus' refers to the general dispositions of social agents or actors, such as their skills, attitudes, values and norms, and so on that do not derive from (specific) roles such as teacher or pupil but from wider experience of life in relation to socio-economic class, gender, ethnicity, religion and so on. In theory, social actors bring to specific social encounters acquired, learned, aspects of self which though not derived from roles involved in the encounter nevertheless partly determine the pattern of interaction that develops among the participants. In other words, *general* life experiences 'affect how we act in any *particular* situation and the dispositional dimension helps us understand'; for example, why, 'no two teachers ever perform the role of teacher in *exactly* the same way'. The third dimension of social interaction in Mouzelis's typology is the *situational-interactional* dimension, which, as Sibeon (ibid.) points out, is described by Mead (1967) as the emergent/contingent features of the interactional situation/encounter itself, and 'in particular to situated meanings that come into existence *during*, and as an outcome of, the process of interaction within a micro setting'

(Sibeon, ibid.) such as a classroom. These are emergent features of social action/interaction derived neither from positions/roles nor from dispositions.

Assessed in terms of usefulness as a post-postmodern analytical device, Mouzelis's typology of dimensions of social action is important in terms of the development of substantive theories and the design of empirical studies.

Derek Layder

As Owen (2009a) suggests, Derek Layder's theories emphasise the relationship between theory and social ontology in respect of methodology (for example, Layder's 1993, and 1998a, work). Even when his primary concern is theory and metatheory (for example, 1994, 1997, 1998b), the question of how theory and method relate to each other runs like a thread through his theoretical work. This point is significant in any analysis of Layder's work. However, in what follows, we will concern ourselves with his metatheorising, particularly his (meta) theory of social domains. Layder, like Sibeon, Archer and Mouzelis, is advocating a 'post-postmodern' renewal of sociological theory and methodology, which is ontologically flexible and avoids the 'absolutist' knowledge-claims of meta-narratives and modernist, *essentialist* paradigms. Layder 'favours a "modest", circumscribed approach to social explanation', which retains an 'epistemological commitment to realism', recognising that 'social reality is multiform, relatively indeterminate and unpredictable' (Sibeon, ibid.: 106–7). This approach is similar to Sibeon's original framework, in that it avoids unitary, *reductionist* explanations for human behaviour. As Sibeon (ibid.: 107) points out, Layder's 'metatheoretical schema' draws upon some contrasting elements and he is critical of some parts of the approaches he draws upon. For example, there is his criticism of the three forms of conflation (downwards, upwards and 'central', which Margaret Archer opposes). Like Sibeon, and as is the case with the framework employed here, and in the work of Owen (2006a, 2006b, 2007a, 2007b), there is an opposition to the idea of duality of structure, indeed towards any attempt to collapse/remove important distinctions between agency and structure, micro and macro and so on. Layder (1998b: 87) criticises symbolic interactionism, Giddensian structuration theory and Foucauldian insights for having what Sibeon (ibid.) calls 'a flattened ontology' which serves to ignore

' "vertical" differentiation of the various spheres (or "domains") of social reality'. In Layder's (ibid.) view, differing forms of phenomena make up the social world (subjective/objective, micro/macro, etc.) and social 'reality' cannot be reduced to unitary, reductionist explication such as 'intersubjectivity', 'habitus', 'figurations' or 'social systems'. Layder appears to favour flexible, multi-factorial explanations as opposed to the relativism of postmodern/poststructuralist accounts.

As Sibeon (ibid.) suggests, Layder's theory of social domains is a largely successful attempt to 'develop a flexible, non-reductionist social ontology as a means of addressing agency-structure and micro-macro'. Layder tends to use several terms interchangeable with 'domains' and 'social domains'. These are 'dimensions', 'orders' or 'layerings', which serve to emphasise the stratified nature of his ontology. Layder contends that no domain has analytical primacy, and that the domains should be viewed as being interwoven and interdependent whilst at the same time autonomous (having distinct properties). Layder's theory of domains suggests that social 'reality' constitutes four social domains, and they relate to the subjective and objective realms. The subjective part of social life has two domains: the individual-subjective referred to as 'psycho-biography', and the intersubjective which is labelled 'situated activity' – the objective/systemic dimension is made of up social settings and contextual resources. The latter term is explained by Sibeon (ibid.) as referred to 'widespread cultural phenomena, and the distribution of resources relating to, for example, social class, gender, ethnicity, and other sources of inequalities and power differences'. Layder maintains that 'objective' (systemic) variables may influence but do not determine the subjective dimension, the latter consisting of psychobiography and situated activity. On the other hand, psychobiography (unique, asocial aspects of the person) and situated activity shape, in part, but do not actually determine social settings and contextual resources (the objective dimension of the social). As Sibeon (ibid.: 108) makes clear:

> *Psychobiography* is the term that Layder uses to refer to largely unique, asocial components of an individual's dispositions, behaviour, and self-identity, these being aspects of the individual that are relatively independent of face-to-face interaction and of

the macro-social sphere. It is worth noting that his concept of psychobiography (Layder 1997) in some aspects is a forerunner of the recent renewal of sociological interest in agency and the self; this interest is reflected in, for example, McNay's (2000) work on gender and agency.

Layder's emphasis upon psychobiography's relative autonomy leads him to repudiate postmodern/poststructuralist 'decentering of the subject', and the idea that social actors are the effects of discourse. He (1997) argues that psychobiography influences the manner in which discourses are 'handled' by individual agents. In other words, psychobiography 'mediates'. Layder is in general sympathetic to the symbolic interactionism of those such as Goffman. However, he criticises the tendency within interactionist sociology to fail to recognise the significance of psychobiography/subjectivity, which is not reducible to interaction/intersubjectivity. He (ibid.) has criticised Elias for a failure to recognise the unique, asocial aspects of individual social actors, and Bourdieu's emphasis upon the type of dispositions (social) acquired through social group membership/experience within his theory of 'habitus' is singled out as 'incomplete' because of the lack of attention given to the psychological dispositions, pertaining to psychobiography, of individual social actors.

The second domain forming part of the subjective dimension of social 'reality' is termed 'situated activity' by Layder and refers to face-to-face social interaction as focused upon in literature of the symbolic interactionist school. As Sibeon (ibid.: 108) suggests, 'Following Goffman (1983: 4), Layder (1997: 88, 93) sees situated activity in terms of situations of co-presence where two or more individuals are able to monitor and reflectively respond to unfolding action.' This would, according to Sibeon, exclude 'larger gatherings of the kind, for example, that contributed to the overthrow of Communist governments in Eastern Europe in 1989/90'. Layder (1997: 133) has suggested that situated activity subsumes 'transient', 'intermittent' and 'regularised' encounters. As Sibeon (ibid.: 109) makes clear, '[t]ransient, one-off encounters include encounters between strangers in public places'; the expression 'intermittent encounters' describes 'for example, meetings between spatially dispersed acquaintances, friends, or family members who periodically make contact with each other'; and 'regularised encounters' are 'ongoing interactions among, say, family members or work colleagues'. Layder (ibid.: 245) describes as a 'hotbed of creativity' the domain of situated activity which forms part of Goffman's 'interaction order'. He suggests that the innovative 'meanings' that frequently emerge in situations of co-presence tend to remain within the location in which they arise. That is to say that in non-routine situations new, emergent meanings may 'travel' across time–space and have implications for the macro-social order, but in routine situations, argues Layder, the new meanings that emerge in locales generally have no implications for other locales or for the macro-social order.

Layder appears to be referring to the home, the street, workplaces and so on when he uses the term 'social settings' (together with *contextual resources* the term forms a dimension of the objective realm of the social). As Sibeon (ibid.) suggests, in Layder's (1998a) scheme, situated activity may 'intertwine' with settings, but settings are actually relatively autonomous. Settings are variable to the extent that they are formally organised, and some settings exhibit highly formalised rules, practices and authority relations whilst others such as family settings are less formalised. All social settings, Layder notes (1997: 3), are 'underpinned by an elaborate social fabric of rules, understandings, objections and expectations'. It could be argued that Layder's work emphasises the importance of differentiating situated activity from social settings, the latter being *locales* of activity (Layder, ibid.: 157). So, social settings may be regarded as locations of activity, relatively autonomous and of the macro-social order. As Sibeon (ibid.: 109) suggests:

> In other words, within settings we encounter conditions of action-discourses, resource patterns, social positions/roles, and more or less institutionalised practices – that are inherited from the past and which, though domains interpenetrate and influence each other, are not reducible to any of the other domains.

The social domain referred to as *contextual resources* by Layder (1997: 4) comprises two elements. As Sibeon points out (2004: 110), firstly this refers to 'society wide' distributions of resources such as 'money, homes, material possessions, health care, and the like', associated with socio-economic divisions of class, gender and race; and secondly, 'widespread cultural understandings' and discourses and social practices. So, in other words, Derek Layder considers *contextual resources* to constitute a domain associated with macro-social patterns of power, domination and material inequality in addition to more widely distributed/patterned 'cultural phenomena', discourses and so on.

Sibeon (ibid.) makes some interesting observations about Layder's social domains theory in the light of issues pertaining to metatheory, ontological flexibility and time–space, for example. These are all concerns which are relevant to the development of the framework here, and have been mentioned in previous discussions. Firstly, Sibeon (ibid.) draws attention to the possibility of contingent links between domains. He argues:

> While each domain is relatively independent of the others, the domains interpenetrate and overlap. There is no simple correspondence or causal connection between domains (for example, interaction in local settings may, depending on the circumstances, tend to subvert or support the macro-social order): in other words, social investigators should be alert to the development of contingent – as opposed to structurally predetermined – links between domains.

Secondly, there is the case that Layder's ontologically flexible approach is empirically 'open' and accords no primacy in analysis towards any of the social domains. Thirdly, Layder's metatheoretical framework can be seen as advantageous in that it incorporates subjective and objective dimensions of the social. This is important because the social ontology has effectively avoided being incomplete/lopsided by this inclusion. Fourthly, Layder has, according to Sibeon, effectively theorised the concepts of time–space. Therefore, as Layder (1997: 77) himself has commented, it is not possible to adequately represent his domains theory, in 'diagrammatic form'. As Sibeon (ibid.: 110) explains:

> We can think of social processes over time as existing in a 'horizontal' plane, but this obscures the 'vertical' layering of the four domains, and even more complexly, such layering is itself dynamic and always in 'process' (1997: 24). Also, there exist differing time frames. Interconnections of objective and subjective

aspects of social reality, and of agency-structure, are complex conjunctures of time-space where the relatively short time-frames of face-to-face interaction (situated activity) meet the extended time-frames of long-standing institutional conditions that extend from the past into the present.

Fifth, Sibeon (ibid.) refers to Layder's stratified social ontology and 'conception of relatively autonomous but interpenetrating domains', stating that the approach contrasts with that of Nicos Mouzelis (discussed previously), who is 'close to Durkheim', when he asserts that macro-social facts must be explained in terms of other macro-social facts, Layder's ontology causes him to reject the idea of explaining macro-social phenomena in terms of other macro-social phenomena, and Layder gives more emphasis to 'the notion of a stratified social ontology in which social phenomena may legitimately be partly explained in terms of other phenomena that exist at differing levels of reality' (Sibeon, ibid.: 111). Sixth, as Sibeon (ibid.) shows, Layder argues that 'there is no *direct link* between action and structure'. As Sibeon correctly asserts, this relates to the theoretical debate around duality versus dualism. Layder (1997: 99, 236) suggests that there is a 'loose coupling' between the relatively independent domains described previously. He has identified (ibid.) the work of Giddens, Bourdieu, Habermas, Berger and Luckmann and Elias, for example, as social theory which, as Sibeon (ibid.) puts it, fails to 'adequately take into account the relative autonomy of the interaction order'. Layder's view (ibid.) is that system reproduction depends upon routine, and what Sibeon (ibid.) refers to as 'reproductive activities in situations of co-presence'. With regard to the aspects of agency–structure discussed here, there is a general agreement between Sibeon, Archer and Layder to the extent that we need to recognise subjective/objective dimensions of society, how social conditions are dependent upon time/space and have emergent properties, and we need to incorporate these insights into social explanation.

In summary, the theorists (Archer, Mouzelis and Layder) discussed are leading theorists whose work is of importance to the development of post-postmodern sociological and crimnological theory. As Sibeon (ibid.: 112) suggests, in relation to Archer's morphogenetic theories:

> There is considerable merit in Archer's conceptual framework, which in large part is built upon critique of three forms

of conflation: downward conflation (which is associated with methodological collectivism) mistakenly attempts to explain agency in terms of structure; upward conflation, which Archer identifies with methodological individualism, interactionism, and phenomenology, is a defective form of reasoning that treats current structures as products of present as distinct from post activities, and assumes structure is no more than 'other actors' and their activities; central conflation, which Archer associates with, in particular, Giddens's theory of structuration, is rejected by Archer for its commitment to a duality of structure and action rather than a dualism in which there is a temporal separation between structure and the action that reproduces or elaborates it.

The author draws attention to other important aspects of Archer's work, including her flexible, 'realist' social ontology. However, Sibeon (ibid.) views Archer's 'conception of the activity dependence of social structure' as unfortunate because it pushes her rejection of upward conflation 'too far', to the point of claiming that some current structures may be entirely independent of current activities.

We have examined the contribution of Nicos Mouzelis towards a sociological understanding of agency–structure and micro–macro, in particular his typology, which is influenced by Durkheim, Parsons, Bourdieu and Mead, and identifies three dimensions of social action: positional/role, dispositional and situation-interactional. As Sibeon (ibid.) suggests:

> This schema-like Mouzelis's insightful elaboration of the system and social integration distinction is developed by Mouzelis into an invaluable analytical tool. Also, its usefulness is capable of being enhanced by the inclusion of social ('organisational') actors in addition to individual human actors, and by including an explicit focus on time-space and material diffusion processes.

However, as Sibeon (ibid.: 112–13) points out, Mouzelis's contribution to the theoretical debates around dualism versus duality is a mixture, on the 'positive' side of 'thought-provoking and potentially illuminating insights', and on the 'negative' side; 'flawed (meta) theoretical interpretations', such as his definition of duality as 'an unreflexive actor orientation to structure', and dualism as 'reflective actor orientation to structure' alongside retaining Giddens's concept of structure yet failing to regard dualism as a 'temporal separation between structure and action and duality as a collapsing of this separation'. Sibeon (ibid.: 113) makes the interesting point that Giddens has noted Mouzelis's misunderstanding of his concept of 'duality of structure'. As Sibeon (ibid.: 113) points out:

> Unlike Archer and Layder, Mouzelis's thesis is that duality and dualism (as defined by him) both have a part to play in social analysis: this postulate is combined with a virtual/actual distinction that refers to paradigm and syntagm, to produce a fourfold typology that encompasses analytical categories that Mouzelis identifies as paradigmatic duality, paradigmatic dualism, syntagmatic duality, and syntagmatic dualism. These formulations by Mouzelis, though they point to interesting questions, are ultimately unsuccessful – as they stand, they serve no useful (meta) theoretical or methodological purpose.

We have examined the contribution of Derek Layder towards our understanding of agency–structure, micro–macro and metatheory. We have also seen how two of the domains, 'psychobiography' and 'situated activity', constitute the subjective realm of the social world, and 'social settings' and 'contextual resources' refer to macro-distributions and ownership of resources, together with widespread discourses and practices. We have also noted that Layder argues against causal links/correspondence between social domains. Any 'links' are likely to be in the form of contingent 'loose couplings'. The four social domains of Layder posses lateral and vertical dimensions, and are complex conjunctions of time–space.

This chapter has outlined recent transitions in social and criminological theory which include a movement away from postmodern/poststructuralist insights towards metatheory. We have examined the concepts and illegitimate forms of reasoning which Sibeon's original anti-reductionism focuses upon and arises out of respectively. We have examined problems with the 'cultural turn' and investigated the work of three major contributors to post-postmodern theory. It has been established that, in order to develop a 'sensitising' framework for the purpose of studying crime and criminal behaviour, Sibeon's original framework provides a viable base with considerable explanatory potential. However, as is shown in Chapter 3, it is necessary to modify Sibeon's original framework to include a focus upon *the biological variable*: the evidence for, at least in part, biological causality in respect of some human behaviour. Evidence for this is provided from evolutionary psychology. We also discuss selected examples of attempts to 'build bridges' between the social and life sciences, for example, writers of the 'embodied' school of sociology.

3
Constructing a Genetic–Social Framework

In what follows, the case is made for an ontologically flexible, metatheoretical framework for the study of crime and criminal behaviour. This would entail focusing above all upon *agency–structure, micro–macro, time–space,* modified notions of Foucauldian *power* and *the biological variable*. The previous chapter discussed agency–structure, micro–macro and time–space (the original concepts which Sibeon's 'sensitising' framework focuses upon), and their explanatory potential. Additionally, we discussed the importance of modified notions of Foucauldian *power*, emphasising that it is important to recognise 'the *dialectical* relationship between *agentic* and *systemic* forms of power; the *relational, contingent* and *emergent* dimensions of power, and the concept that *contra* Foucault, aspects of power can be stored' in positions/roles and in social systems/networks (Owen, 2012a). This chapter includes arguments for the modification of Sibeon's original framework to include a focus upon *the biological variable*: the evidence from evolutionary psychology (EP) and behavioural genetics for a, at least in part, biological basis for some human behaviour. The chapter also explores further ideas pertaining to dualism between the social and biology as opposed to duality, and looks at the attempts of those who seek to redefine the relationship between 'nature' and 'the social'. This includes writers of the 'embodied' school. The argument for the inclusion of 'extra' cardinal sins (alongside Sibeon's original *anti-reductionist* concepts of *reductionism, reification, essentialism* and *functional teleology*) such as *'the oversocialised gaze'* and *'genetic fatalism'* is also reinforced in this chapter. In what follows, we briefly codify the framework which is applied in Chapter 4 to selected aspects of crime and criminal behaviour. Later, we look at the possibilities of 'building bridges' between the social and life sciences, and at the possible need to acknowledge *the biological variable*, neglected by Sibeon's original framework.

Codification of the metatheoretical framework

The previous chapter made the case for the post-postmodern explanatory potential of *metatheory*. Briefly, as Sibeon (2004: 13) makes clear, *metatheory* is designed to equip us with 'a general sense of the kinds of things that exist in the social world, and with ways of thinking about the question of how we might "know that world" '. The framework employed here is an example of *metatheory* consisting of methodological as distinct from substantive generalisations. It is argued here that the framework is capable of contributing both to the 'return to' sociological theory and method, and to a post-postmodern criminological theory. The 'pay off' will hopefully lie in the framework's explanatory potential in terms of theoretical development. Rather than being 'hot house debate' it is the author's contention that the framework's usefulness lies in its avoidance of anti-foundational *relativism, oversocialised* (harshly 'environmentalist') accounts; the tendency towards *genetic fatalism* (the equation of genetic predisposition with inevitability); and its adoption of the 'sociological realism' of those such as Mouzelis (1991, 1993a, 1995) who seek a 'return to' sociological theory and methodology. The framework has been employed by Owen (2006a) in a theoretical contribution towards a 'Post-Foucauldian' sociology of ageing – by Owen (2006b, 2009a) in the form of 'Genetic-Social Science' in order to study biotechnology, by Powell and Owen (2005) to critique the 'Bio-medical model' of ageing and recently in briefer studies of crime (Owen, 2007a, 2009, 2012a, 2012b).

The 'sensitising device' arises out of a critique of six forms of illegitimate theoretical reasoning. The first four are found in Sibeon's (1996) original framework (which he called *anti-reductionist sociology*). They are *reductionism, essentialism, reification* and *functional teleology*. To recap, a *reductionist* theory is a theory that attempts to explain social 'reality' in terms of a single, unifying principle such as 'patriarchy' (Hindess, 1986a, 1988). *Essentialism* 'presupposes in aprioristic fashion a necessary unitariness' or "homogeneity" of social phenomena' (Sibeon, 1997b: 1). *Reification* 'involves the illegitimate attribution of agency to entities that are not actors or agents' (ibid.). *Functional teleology* refers to 'illicit attempts to explain the causes of social phenomena in terms of their effects' (Betts, 1986: 51). The second two, which are employed by Owen (ibid.) and Powell and Owen (ibid.) and here, are *genetic fatalism* and the *Oversocialised Gaze*. *genetic fatalism* refers to the widespread tendency within social science to associate genetic predisposition with inevitability (Ridley, 1999, 2003). *The Oversocialised Gaze* refers to 'Environmentalist' accounts characterised by a strong antipathy towards genetic or even partially genetic explanations, such as Giddens's (1993: 57) suggestion that 'human beings have no *instincts* in the sense of complex patterns of unlearned behaviour'. Aside from focusing upon *agency–structure*, *micro–macro* and *time–space* (included in Sibeon's original framework and explored/defined in the previous chapter), the 'new' framework also entails focusing upon the *biological variable*: the evidence from evolutionary psychology and behavioural science for a, at least in part, genetic basis for some human behaviour (Ridley, 1999, 2003).

'Building Bridges'

In what follows, we examine the work of several authors who have attempted to straddle the biology–social divide and consider whether it is possible to move beyond the established 'starting point' for a *biological sociology* outlined in the work of Shilling (1993). The term 'genetic-social' is employed for the framework employed here, as the intention is to distance it from the sociological weaknesses of many so-called biosocial explanations of crime and criminal behaviour (see, for instance, Walsh and Beaver, 2009; Walsh and Ellis, 2003), which, although dealing adequately with biological variables, appear to neglect or make insufficient use of meta-concepts such as agency–structure, micro–macro and time–space in their accounts of the person. Among authors who have attempted to 'cross the divide' between the social and life sciences, Layder's (1993, 1997) work incorporates the interesting and useful concept of *psychobiography* (discussed in terms of explanatory potential in the previous chapter) – the largely unique, asocial components of disposition, behaviour and self-identity. These aspects are regarded by Layder as relatively independent of face-to-face interaction. For Layder, human beings are composed of unique elements of cognition, emotion and behaviour that are, in some sense, separable from the social world, whilst at the same time related in various ways to social conditions and social experiences. Layder is critical of postmodern attempts to 'de-centre' the subject, 'the cultural turn' in general, and symbolic interactionism. The latter school of sociology is singled out for particular criticism because of its tendency to emphasise intersubjectivity to an extent that largely ignores individual subjectivity (or *psychobiography*).

There are possibly philosophical and sociological connections between the psychobiographical components of Layder's 'Social Domains' theories and the work of Benton (1991, 1993, 1994, 1999, 2003), and Hochschild (1983, 1990). Arguably, all three authors have managed to straddle the biology–social divide in a way that so many sociological and criminological accounts have failed to do. Hochschild's work on emotion, as is the case with the work of Taylor (2006), stands in sharp contrast to 'social constructionist' models of emotion. Such 'mainstream' sociological accounts tend to prioritise the biological and the social respectively. Hochschild (1983) adopts a contrasting position which occupies an analytical space between the two extremes. For Hochschild, emotion is unique among the human senses, related to action and cognition. This is similar to Layder's views pertaining to unique, psychobiographical aspects of the person. Emotion emerges, according to Hochschild, when bodily sensations are joined with what we see or imagine, and on this basis we discover our own particular views of the world, and our readiness to act within it. Such an approach serves as a valuable lesson on the role of *the biological variable* in sociological explication. For a considerable time, what Freese et al. (2003) call 'biophobia', a hostile reaction on behalf of sociologists and criminologists towards explanations which evoke biology, has been in evidence. The mere mention of biological causality has tended to lead to accusations from 'mainstream' sociology of a vulgar and crude reductionism. As Hochschild (ibid.) argues, whilst a sociology of emotions needs to 'go beyond' the biological, this does not mean leaving it out altogether. Conversely, if we choose to incorporate biology into sociological explanation, this does not necessitate an overly deterministic role for the latter. Cases in point are the work of Owen (2006a, 2006b, 2009, 2012a, 2012b),

Powell and Owen (2005) and Quilley and Loyal (2005). In the latter example, it is argued that Elias's work provides a 'central theory', serving to integrate diverse theoretical traditions within sociology, whilst providing a framework for establishing a synthesis across the full range of social and biological human sciences. Any sociological/criminological enquiry into human emotions will surely confront 'the limits of the social', but it is the case that human biology displays a degree of plasticity in relation to wider socio-cultural influences (Ridley, 1999: 306). For example, feelings are 'not stored' inside human social actors, but the management of feeling may contribute to the creation of it. Here Hochschild has, arguably, developed a fairly sophisticated perspective on human emotions which side-steps crude, reductionist debates.

Benton's (1991, 1994, 1999) work criticises the 'idealist' tendency in contemporary social theory and the postmodern 'cultural turn'. He suggests that extreme relativism in the philosophy and sociology of science has interbred with a 'pervasive misreading of the Saussurean thesis of the arbitrariness of the relation between sign and signified' and also that 'Nietzschean moral and epistemological nihilism' has produced 'luxuriant growths of hyper-idealism' (Benton, 1994: 45). The author also suggests that the complex network of 'categorical oppositions (mind/body; culture/nature: society/biology, meaning/cause, human/animal) which structure the discourses and define the research agenda of these prevailing traditions now constitute intellectual obstacles' in the way of meeting the challenge to 'settled' conceptual forms in 'mainstream' sociology (ibid.: 25). Benton considers that his work has provided a preliminary blueprint for how 'dualistic', anti-naturalistic programmes in the contemporary human sciences should be understood. As it is made clear in the previous chapter, the framework employed here favours *dualism* as opposed to a d*uality of structure*. As it is explained later in this chapter, this is in order not only to avoid the compacting of agency and structure but also to avoid the collapsing of distinctions between the 'biological' and 'the social'. According to Benton (ibid.), 'dualistic' programmes are best understood as primarily defensive reactions to 'reductionist' disciplines such as Social Darwinism. The author's view is that the 'task' for any re-alignment of sociology and related social sciences with the life sciences is one of providing conceptual room for organic, bodily and environmental aspects, and aspects of social life, to be given their 'proper place' without discarding the achievements of classical social science in defense of the autonomy of those disciplines vis-à-vis the life-science specialisms. Importantly, like Hochschild, Benton (ibid.) suggests that a realignment or even integration between disciplinary fields such as biology and criminology does not have to mean reductionism, or a reduction of the one to the other.

Recently, as Owen (2006a) suggests, Armstrong (1987: 1217), drawing upon Foucauldian insights, has warned against too readily endorsing 'an alliance of bio-social perspectives'. Similarly, a range of Foucauldian arguments have emerged, claiming that medical power should be seen as a 'dangerous' form of power and surveillance (Biggs and Powell, 1999; Katz, 1997; Powell and Biggs, 2000). These perspectives go some way towards offering *social* understandings of disease that may be juxtaposed against the biological definitions operant in hospitals and other fields. Scepticism towards biological 'determinism' can also be found in the Foucauldian analysis of Powell and Biggs (2004: 2), in which it is argued:

> A significant contribution of sociology as a discipline has been to highlight how individual lives and illnesses which were thought to be determined solely by biological, medical and psychological factors, are, in fact heavily influenced by social environments in which people live.

The authors are arguably correct to suggest that social variables play a large part in 'influencing' behaviour, lives and illnesses. As Ridley (1999: 306) suggests, social variables are probably in total 'more important than genes in nearly all behaviours'. However, it is argued by Owen (2006a, 2006b, 2007a, 2007b, 2009, 2012a, 2012b) that we should avoid associating genetic predisposition with inevitability. As previously stated, the framework employed here arises out of a critique of several illegitimate forms of reasoning, one of which is *genetic fatalism*. According to Ridley (ibid.), the tendency to equate biological predisposition with inevitability is fairly widespread in social science.

Newton's (2003: 35) work pertaining to the possibility of 'marriage' between the social and biology suggests that the human body can be 'both extra-discursive and deeply embedded in the social fabric' as can be observed in the human smile, body scripts of birth, maturation and death and so on. In his view, such examples serve to illustrate the *social* reliance of the 'extra-discursive body' and how any sociological account of the social world is deficient if it 'ignores the fact that human beings have *biological* bodies' (ibid.). According to him, the significance of the human body for the social serves to emphasise the need to formulate concepts capable of integrating the two. However, he highlights some of the problems facing those such as Benton (1991), Bury (1995, 1997) and Williams (1998) who are attempting to cross the 'Great Divide' in the sense of furthering the project of a material-corporeal sociology. In particular, Newton draws attention to the reliance on animal studies in order to explain human health (Freund, 1998: 854–5, 1990: 464, 1998: 277) and Williams (1998: 129, 2003: 71), arguing that 'reference to animal studies incites a reductionist account of human health and appropriates rather inappropriate analytic concepts-features which appear to contradict Freund and Williams' expressed concerns' (ibid.: 28). Newton sympathises with Shilling's (1993) attempts to analyse the human body as *simultaneously* biological and social, viewing this 'Third Way' as a starting point for a *biological sociology* which goes beyond naturalistic and social constructionist views of the human body, whilst retaining some of their useful insights. However, he illustrates how there are serious methodological and epistemological problems facing current attempts to move beyond Shilling's (ibid.) starting point. He suggests that a strong desire to counter-constructionism may lead to an 'insensitivity' towards constructionist arguments, citing the work of Freund and Williams again as examples of authors who lack a 'reflexivity to their own *emotional* evocation of the supposed links between the social (e.g. inequality), the emotional (and its physiology) and ill-health' (Newton, ibid.: 36).

In sum, Newton urges us to remember that we remain 'at the starting point' of a marriage between the social and the biological. In his view, we cannot rely upon current psycho-physiological research in order to conceptualise biology, and we face further problems in relation to the complexity of the relationship between the human body and the social world. Indeed, in his view, we cannot necessarily interpret a particular physiological measure as a 'response' to social stimuli. For Newton, 'neither the social nor the biological generally operates in this singular fashion' (ibid.). He also draws attention to the difficulties of isolating 'singular relationships' within the biological field, characterised as it is by great complexity.

It is argued by Owen (2006a, 2006b, 2007a, 2009, 2012a) that we *can* move beyond Shilling's (ibid.) 'starting point' for a marriage between biology and the social. We can incorporate the concept of *genetic fatalism* (as a 'cardinal sin') into a 'new' framework, and focus upon *biological variables* (evidence for genetic causality). Arguably, the greatest error in the work of many of the authors who have attempted to straddle the biology–social divide (in particular, Benton, 1991, 2003; Newton, ibid.; Shilling, ibid.) appears to lie in the tendency not to equate predisposition with genes but to equate predisposition with *inevitability*. As Ridley (1999: 307) clearly states:

> Suppose you are ill, but you reason that there is no point in calling the doctor because either you will recover, or you won't: in either case, a doctor is superfluous. But this overlooks the possibility that your recovery or lack thereof could be caused by your calling the doctor, or failure to do so. It follows that determinism implies nothing about what you can or cannot do. Determinism looks backwards to the causes of the present state, not forward to the consequences.

In the 1970s, after the publication of E.O. Wilson's book *Sociobiology*, there was a counter-attack against the idea of genetic influences upon human behaviour, led by Richard Lewontin and Stephen J. Gould (Wilson's Harvard colleagues). Their dogmatic slogan 'Not in our genes!' may have been a plausible hypothesis at the time, but as Ridley (ibid.: 306) argues, 'After 25 years of studies in behavioural genetics, that view is no longer tolerable. Genes do influence behaviour.' Perhaps we may move beyond Shilling's (ibid.) suggestion that the human body is *simultaneously* biological and social, by acknowledging Ridley's (ibid.) point about genetic influences. We should perhaps recognise that, even after the genetic discoveries, Ridley mentions, environmental influences remain enormously important. In the light of Newton's (ibid.) call for a 'more sophisticated' understanding of biology and psychology, perhaps we should strive to avoid sociological and criminological accounts which deny the possibility that the physical can be 'at the mercy' of the social, such as those of Armstrong (ibid.). As Ridley (ibid.: 153) suggests:

> If genes are involved in behaviour then it is they that are the cause and they that are deemed immutable. This is a mistake made not just by genetic determinists, but by their vociferous opponents, the people who say behaviour is 'not in the genes', the people who deplore the fatalism and predestination implied, they say, by behaviour genetics. They give too much ground to their opponents by allowing this assumption to stand, for they tacitly admit that if genes are involved at all, then they are at the top of the hierarchy. They forget that genes need to be switched on, and external events – or free-willed behaviour – can switch on genes.

There is a growing, emerging literature on the subject of the possibility of a 'marriage' between the biological and social sciences. So far we have examined the work of several theorists of the 'embodied' school, and a 'voice' from evolutionary psychology in the form of Ridley. It is important to widen and deepen the focus to include an examination of more diverse sources, and to tentatively apply some of the concepts from the 'new' framework to some of the theoretical positions in the debate (for example, the 'embodied school'). Chapter 4 deals with a more in-depth application of the framework to the study of selected examples of crime and criminal behaviour. For the moment, we will examine further the literature pertaining to links between the social and biology, and consider whether it is viable, given the evidence, to include the *biological variable* alongside *agency–structure, micro–macro* and *time–space*.

Williams (2003: 550), another writer associated with the 'embodied school', or 'the sociology of the body' makes clear his position within the debates around the possibility of 'marriage' between the social and the biological, and the broader context within which the debates are located. The author refers to the current 'state' of the debates within sociology and describes several 'schools of thought'. Alongside those such as Dickens (2000) who attempt to develop an 'alternative kind of social Darwinism' (Williams, ibid.: 556), there are those such as Williams himself, Freund, Benton and Shilling who may be located within 'embodied sociology'. In Williams' view, it is the 'embodied' theorists who have been particularly important in 'challenging if not dispensing with dualist legacies of the mind/body, reason/emotion, biology/society kind' (Williams, ibid.: 550). Also referred to by the author are the recent developments within *the sociology of health inequalities*, such as the 'psychosocial perspective' of Wilkinson (1996, 2000a, 2000b) which represents something of a departure from 'embodied sociology'. As Williams (ibid.) explains, for Wilkinson (ibid.), health is 'no longer limited primarily by the direct effects of material factors in rich societies that have gone through the "epidemiological transition"'. Elstad (1998: 600) has commented that there are three central contentions with the *psychosocial* approach, which are '(1) the distribution of psychological stress is an important determinant of health inequalities in present-day affluent societies, (2) psychological stress is strongly influenced by the quality of social and interpersonal relations, and (3) the latter are determined to a large extent by *the magnitude of society's inequalities*'. As Williams (ibid.: 552) points out, there are four main sources for the psychosocial perspectives of Wilkinson (2000a). These are the *social stress* approach, the related *self-efficacy* approach, the newer *sociology of emotion* and the *social cohesion* approach.

For Elstad (ibid.: 601), it is not a case of 'whether *some* health-related bodily changes may follow from mental appreciations of external circumstances'. In the author's view, recent reviews leave '*few doubts that psychological stress, generated by despairing circumstances, insurmountable tasks, or lack of social support, can influence disease-related patterns*'. The author identifies examples such as 'the endocrine system (eg secretions of catecholamine and cortisol), and the immune system (eg secretions of catecholamine and cortisol), and the immune system (eg number of T-cells and Natural Killer cells)'. However, the problem is '*to demonstrate that such changes are large enough and long-term enough to affect health in a significant way*'. Newton (2003) raises likewise concerns, as do Wainwright and Calnan (2002). However, Williams (ibid.: 553) considers the idea that we should '*question the whole relationship*', premature as 'the pathways are *more than* plausible'.

Williams (ibid.: 555) discusses the 'heated debate' that has been sparked by 'neo-material critiques of psychosocial perspectives or pathways, including important questions concerning the "causes of the causes", the deeper relations of (class) inequalities and the (global) role of neo-liberalism (Coburn, 2000a, 2000b; Scambler,

2000; Wilkinson, 2000b)'. In his view, what is required is the need to think in terms of relationships between, 'social structure, ecological conditions of life, bodily constitution and activity, social and cultural processes, and mental and emotional dispositions, skills and experiences' (Benton, 2003: 285). Williams considers that such a position avoids what Benton (ibid.: 293) has called 'the simplistic opposition of "nature" versus "nurture" in relation to health'.

Interestingly, on the one hand, Williams (ibid.: 556) argues that 'if we want to challenge dualism on all fronts, as "material corporeal sociologists" appear to, then one important route, as Birke (2003) reminds us, must be to take on board the growing calls to recognise at least some kinds of non-human animals as "clever, adaptable, aware and cultural – just like us" '. On the other hand, the author appears to condone dualism when he states '*we need to respect the discrete analytic potential (and autonomy) of various disciplines*, biology, criminology and sociology included, thereby countering the (reductionist) slide from the having of *something* important to say about the world and our place within it, to the having of *everything* to say on these matters' (ibid.: 558). Despite this apparent contradiction in terms, Williams sees 'the way forward' in the biology–society debate as a 'multi-disciplinary' approach which would serve to 'dissolve the dualistic opposition between "society" and "nature" without giving in to either a social constructionist reduction of nature to culture, or the reverse reduction of social life to a mediated epiphenomenon of the human genome (Benton, 2003:292)'.

Evolutionary psychology and 'nature via nurture'

Evidence for the idea that human beings evolved complex behaviour by the addition of instincts to those of their ancestors rather than by replacing instincts with learned patterns of behaviour can be found in the work of the psycho-linguist Pinker (1994). As the work of Owen (2006b, 2012a) shows, Pinker's argument is that the intention of most social scientists has been, and is, to examine and trace the ways in which behaviour is influenced by the social environment. We should perhaps examine instead the ways in which the social environment is produced by our innate social instincts. As Ridley (1999, 2003) has suggested, the 'fact' that all human beings smile at happiness and frown when worried may be expressions of unlearned behaviour (instinct) rather than culture.

The evolutionary psychologists Barkow et al. (1992) suggest that culture is the product of individual human psychology more than vice-versa. Arguably, it has been a major mistake to oppose 'nature' to 'nurture', because learning is, according to the authors, dependant upon innate, instinctive capacities to learn, and there are innate constraints upon what can be learned. As Ridley (1999: 103) makes clear, the 'evolutionary' in evolutionary psychology refers not to natural selection itself but to the concept of adaptation. Biological organs can be 'reverse engineered to discern what they are "designed" to do', in the same way as machines can be so studied. Pinker (ibid.) emphasises that machines are *meaningless* except when described in terms of their functions. In the same way it is meaningless to describe the human or animal eye without mentioning that it is 'designed' for the making of images. Barkow et al. (ibid.) argue that the same principles apply to the human brain, and its modules are likely to have been 'designed' for particular functions. Ironically, as Ridley (ibid.) points out, 'it was the argument from design that kept evolutionary ideas at bay throughout the first half of the nineteenth century'.

'The evidence that grammar is innate is overwhelming and diverse', claims Ridley (ibid.: 104). The author refers to the language instinct possessed by all human beings as a complex and sophisticated adaptation for communication. He conceives how 'it was of great advantage to our African ancestors to share precise and complex information at a level unavailable to any other species'. Ridley considers the evidence that grammar is innate rather than learned to be abundant:

> The evidence that a gene somewhere on chromosome seven usually plays a part in building that instinct in the developing foetus's brain is good, though we have no idea how large a part that gene plays.
>
> (ibid.)

However, many social scientists appear to be firmly resistant to the idea of human instincts. For example, Giddens (1993: 57) suggests that '[h]uman beings have no *instincts* in the sense of complex patterns of unlearned behaviour'. Yet, as Owen (2006b: 96) points out, 'the evidence for instinct, and for genes whose effect seems to be to achieve the direct development of grammar remain strong'. Certainly, the work of Dunbar (2003a, 2003b, 2004), Barrett et al. (2002) and Dunbar et al. (1999) seems to support this position. As Ridley argues (ibid.: 105), it appears to be the case that many social scientists prefer to think, contrary to an abundance of biological evidence, that genetic influences upon human language are 'mere side-effects of its direct effect on the ability of the brain to understand speech'. Arguably, the twentieth-century paradigm to the ends that only non-human animals possess instincts begins to develop cracks in the foundations 'once you consider the Jamesian idea that some instincts cannot develop without learnt, outside inputs' (Ridley, ibid.). The reference here is to William James, the early psychologist who believed that human behaviour revealed evidence of more separate rather than fewer instincts than non-human animal behaviour. Yet the belief that human beings possess no innate characteristics outside of 'a set of simple reflexes, plus a range of organic needs' (Giddens, ibid.) appears to be firmly entrenched within the discipline of sociology in particular.

Despite reservations about Lewontin's (2000) *genetic fatalism*, Owen (2006b: 905) 'recognises that the author has cogently identified biological discourses which are remarkably similar to the dominant ideas of their day'. This point is emphasised because Lewontin is a most prominent example of sociologists who appear to deny biological causality. Lewontin's contention that nineteenth-century 'racist scientists' produced 'racist science' is controversial to say the least, and must be taken seriously. His suggestion that a 'democracy of the elite' operates within science is far from a trifling issue. The implications are that the apparatus of knowledge called 'scientific knowledge', depending upon elaborate metaphors, is available only to a restricted elite. Additionally, his suggestion that there is a 'dark side' to the manner in which the Creationist movement in the US Bible Belt has 'thrown out' all science rather than re-evaluate inherited notions of God importantly identifies a mindset which automatically opposes any form of change to the mythical 'timeless' landscape (Owen, ibid.). Likewise, Lewontin's timely criticisms of the idea of the Human Genome Project as a 'recipe book' for life cannot be dismissed lightly.

Particularly strong criticisms of evolutionary psychology can be found in the work of Midgley (2000). The author argues that evolutionary psychologists 'teach' determinism and nihilism, whilst treating human beings as solitary entities rather than social animals. Daly and Wilson's (1998) work appears to contradict Midgley's point regarding the reliance upon social atomism by evolutionary psychologists. The authors examine step-relationships and acknowledge that they 'exist in all societies' (Daly and Wilson, ibid.: 37–8).

Perhaps at this point, it might be useful to locate evolutionary/psychology within wider sociological debates around 'the new genetics'. David (2002: 303) locates evolutionary psychology's 'mass modernity account of human intelligence' within discourses pertaining to the 'naturalisation of society'. For David, current sociological interest in genetics and biotechnology 'focuses attention on three interrelated fields: the "naturalisation of society" (Benton, 1999; Dupre, 2001); the "socialisation of nature" (Dickens, 1996; Irwin, 2001); and the "normalisation of the individual" (Ettorre, 2000; Petersen and Bunton, 2002)' (ibid.). The 'socialisation of nature' refers to 'attempts to apply genetic technologies to modify non-human nature (plants and animals)' (ibid.). The 'normalisation of the individual' focuses upon 'the application of genetic diagnostics and therapies within the realm of human health and medical practice' (ibid.). General overviews of evolutionary psychology (such as those provided by Barkow et al., 1992; Buss, 1999; Dunbar et al., 2006; Pinker, 1997) tend to emphasise that evolutionary psychology 'argues that the mind is a collection of special-purpose software designed by natural selection to solve the problems of survival and reproduction that faced our ancestors' (Curry, 2003: 1).

As previously emphasised, evolutionary psychology is not without its critics, and our task after examining the literature is to decide whether there is sufficient evidence from the discipline to justify modifying Sibeon's original framework so that it entails focusing upon *biological variables* (evidence for biological/genetic causality, at least in part, with regard to some human behaviour).

David (ibid.: 304) considers evolutionary psychology to be 'only the latest manifestation of biological reductionism', seeking to 'identify the roots of the contemporary human condition in the stone age/s'. He is sceptical about 'reverse engineering' (the attempt to 'infer the ancestral conditions, called the environment of evolutionary adaptation, that would have made certain genetically inherited behaviour-inducing modules increase their bearer's reproductive success') (ibid.), because it is 'empirically untestable' (ibid.: 312). The author identifies the 'existence of hard-wired mental "modules"' designed to produce 'programmed response'. These 'adaptionist' principles imply that 'what exists is said to have evolved' (ibid.: 304). Whilst others such as Rose and Rose (2000) have criticised evolutionary psychology regarding theoretical deficits pertaining to rape, social competition and so on, David concentrates upon addressing the 'core proposition' of the discipline, the 'modularity of mind'. The author feels that sociology must 'challenge reductionist science', of which evolutionary psychology is an example, whilst contributing to a wider understanding of 'what it does mean to be human' (ibid.). David identifies the concept of the modularity of the human mind as a central concern in the work of several leading evolutionary psychologists, such as Barkow et al. (1992), Pinker (1995, 1999) and Gigerenzer et al. (1999). The basic idea is that the mind is 'made up' of a series of specific modules adapted to provide 'functional responses' (David, ibid.) to specific stimuli/problems, rather than being a 'single processor able to use general reasoning principles or mechanisms to solve problems' (ibid.). Therefore, in David's view, this means that evolutionary psychologists may apply the principles of natural selection to psychological phenomena. The module 'by analogy with the gene' (ibid.) becomes the unit of selection. Referring to Fodor's (1983) formulation of modularity, which assumes a 'dual processor model', David makes the point that 'far more of our *thinking* is "preprogrammed" than is assumed in dual processing theory'. However, the evidence for this claim appears to be lacking. The author argues that the general programme of evolutionary psychology is to 'extend the logic of "hard-wired" mass modularity to every aspect of human behaviour' (ibid.).

Referring to the common metaphor given for evolutionary psychology's concept of mass modularity, the Swiss Army Knife of Barkow et al. (1992), David (ibid.: 305) considers such an account (in which the human brain/mind is conceptualised as a 'multiple of relatively independent blades, each one built to carry out a specific task') to be a 'purpose-built intellectual module serving a singular function', which is to deny the 'possibility that humans could be other than they are'. The author adds that 'such a model of the mind fits very nicely into the current intellectual and social climate'. He goes on to quote Hilary Rose (2000: 107) who argues that 'a restored and re-energised neoliberalism has provided the perfect ecological niche for a new wave of biology as destiny'. Interestingly, Curry (2003: 9) in a review of Hilary Rose and Steven Rose's (2000) *Alas, Poor Darwin: Arguments against evolutionary psychology*, suggests that the Roses have not provided evidence that the ideology is motivated by 'unexamined political presuppositions'. David's (ibid.) claim to the ends that a neatness of fit between the model of the mind favoured by evolutionary psychologists and current intellectual and social fashions enables the outlook to flourish is arguably contestable. David (ibid.) refers to evolutionary psychology advocates (though he does not provide an example) who suggest that 'rationality cannot be defined outside its reproductive functionality', and 'if a way of *thinking* helps an organism survive and reproduce it is said to be rational'. As evolutionary psychology is well adapted in the political environment it therefore displays evolutionary rationality according to the author. David (ibid.) goes as far as to say that in its own terms EP therefore displays evolutionary rationality. Aside from the lack of a referenced example, David arguably engages in *reification* here (one of the 'cardinal sins' of the framework), and refers to evolutionary psychology as if it is a social actor possessing the agency to formulate and act upon decisions.

David (ibid.) does, however, provide some evidence from the work of Karmiloff-Smith (1992, 2000) which may demonstrate some weaknesses in the theory of mass modularity. According to David, Karmiloff-Smith (ibid.) provides evidence that studies of brain damage taken by evolutionary psychologists to demonstrate modularity actually demonstrate the '*plasticity* of the brain' (David, ibid.). Damaged brains show forms of wiring which are different from those of non-damaged brains and not only in 'the dysfunctional areas' (ibid.). There is evidence, it is claimed, that the human brain 'appears to compensate for damaged regions by creating alternative neural connections' (ibid.). As David suggests, 'far from proving full modularity, such damage limitation suggests that, while there is a degree of modularity, the brain actively modifies any blueprint it may be built upon to limit general dysfunction' (ibid.). Essentially, David appears to be suggesting that Karmiloff-Smith's work contains evidence that there is a less 'fixed' relationship between 'region' and 'function' within the human brain than is implied by the evolutionary psychological notion of modularity. David (ibid.: 304) refers to Fodor's (1983) original formulation of mass modularity, and has, as we have seen, used Karmiloff-Smith's (1992, 2000) work to identify possible deficits in the theory. However, Fodor (2001) published *The Mind Doesn't Work That Way*, which effectively dismantles his original formulation of mass modularity. There is no mention of Fodor's later work in David's (2002) paper. To be fair to Fodor (ibid.), Ridley (ibid.: 66) shows how the author argues that 'breaking down the mind into separate computational modules was by far the best theory around' (and David (ibid.) does acknowledge 'a degree of modularity'), 'it did not and could not explain how the mind works'. According to Ridley (ibid.), Fodor (2001) is arguing that:

> Minds are capable of abducting global inferences from the information supplied by the parts of the brain. You may see, feel and hear raindrops with three different brain modules linked to different senses, but somewhere in your brain resides the inference: 'it is raining'. In some inevitable sense, then, thinking is a general activity that integrates vision, language, empathy and other modules: mechanisms that operate as modules presuppose mechanisms that don't. And almost nothing is known about the mechanisms that are not modular. Fodor's conclusion was to remind scientists just how much ignorance they had discovered: they had merely thrown some light on how much dark there was.

David (ibid.: 309) refers to 'genetic reductionist accounts of the mind' which draw upon 'misleading analogies with computers'. Citing the work of Pinker (1999), David refers to the tendency, as he perceives it, in evolutionary psychology to view intelligence as 'computationally modular'. This model is challenged by those in theoretical sociology such as Collins and Kusch (1998), and those in neuroscience such as Van Gelder (1995). Collins and Kusch (ibid.) are said by David (ibid.) to have developed an analytical framework for distinguishing between types of action. In theory, the frame enables greater insight into differences between the capabilities of human beings and machines such as computers. The authors highlight the 'higher-order intelligence which most human "work" requires, intelligence which machines are still capable of achieving' (David, ibid.). Collins and Kusch (ibid.) distinguish between action and behaviour, with action seen as 'intentional' rather than reflex/mechanical. Within the action category, as David (ibid.) shows, Collins and Kusch (ibid.) distinguish between polimorphic and mimeomorphic forms of action. 'Mimeomorphic action is action that can be specified fully in terms of its behavioural components (and the situations in which it is appropriate) without an understanding of the "form of life" of which that action is a part' (David, ibid.). These then, are intentional actions. Polimorphic actions are 'those where the capacity to successfully follow the rules of the form of life in which the action takes place is crucial for the correct performance of the action' (ibid.). Collins and Kusch (ibid.) give the examples of 'writing love letters' and 'voting' as acts which cannot be mechanised, as according to David's (ibid.: 310) account of their work 'the act is not the mechanics, but the expression of an intention requiring the capacity to participate in, at however simple a level, the political life of a community'. Polimorphic actions are 'higher-order' forms of action 'requiring the capacity to follow cultural rules that are themselves context-dependent and which, therefore, cannot be specified outside the cultural practices of which they are a part' (David, ibid.). For Collins and Kusch, forms of 'playful' polimorphic action include human attempts to 'enact the same behaviour to relieve boredom or express individuality' (David, ibid.) as here intention rather than the capacity to mimic behaviour is manifested. 'It is polimorphic also to the extent that rule bending is itself a social institution with culturally enacted rules that must be interpreted correctly, in the eyes of other cultural players, to be carried out successfully', suggests David (ibid.). Collins and Kusch (ibid.) are, according to David (ibid.), firmly in favour of the concept that polimorphic actions are not modular in themselves because the rules they enact are unable to be specified in a modular way. David's critique of evolutionary psychology appears to treat evolutionary psychology as 'only the latest manifestation of biological reductionism' (ibid.: 304). However, as Ridley (2003: 66) shows, within the 'school of thought', there are subtleties which those such as David (2002) appear to miss. For example, with reference to computation in the human brain, Ridley (ibid.: 66) suggests that:

> To build a brain with instinctive abilities, the Genome Ordering Device lays down separate circuits with suitable internal patterns that allow them to carry out suitable computations, then links

them with appropriate inputs from the senses. In the case of a digger wasp or a cuckoo, such modules may have to 'get their behaviour right' first time and may be comparatively indifferent to experience. But in the case of the human mind, almost all such instinctive modules are designed to be modified by experience. Some adapt continuously throughout life, some change rapidly with experience then set like cement. A few just develop to their own timetable.

Far from being the 'genetic reductionist accounts' which David (ibid.: 309) refers to, it would appear that evolutionary psychology as a discipline is capable of acknowledging a degree of plasticity of the brain alongside modularity (Fodor, 2001; Ridley, 1999, 2003), and that instinctive modules of the brain are 'designed' to be modified by experience and environmental stimuli (Ridley, 2003). Indeed, Ridley (ibid.: 280) has made the point that 'nature versus nurture is dead. Long live nature via nurture'.

Reference was previously made to the work of Hilary Rose and Steve Rose. They are particularly harsh critics of evolutionary psychology, and have published several papers and books (later referred to here) which arguably accuse evolutionary psychologists of producing work which is reductionist, determinist and adaptationist (Curry, 2003). It could also be argued that their *Alas, Poor Darwin: Arguments against Evolutionary Psychology,* published in 2000 is perhaps their most in-depth critique. At this point it should be firmly emphasised that both Steve Rose and Hilary Rose have unassailable reputations as scholars, and that they have both contributed greatly to the literature on evolutionary psychology. It is not the intention here to suggest otherwise, merely to cover Curry's criticisms of their respective positions on evolutionary psychology. Curry (2003: 2) suggests that the collection of essays brings critics together to 'argue that evolutionary psychology is a "fashionable ideology" whose adherents are "fundamentalists" who promote "simple minded", "socially irresponsible", "culturally pernicious" explanations of human behaviour that rest on "shaky empirical evidence, flawed premises and unexamined political presuppositions"'. In what follows, we will briefly consider the main premises of evolutionary psychology and the Roses' five main arguments against it together with their account of the politics of the discipline, as outlined by Curry (ibid.).

According to Dawkins (1976), modern evolutionary biology revolves around the idea that genes build organisms in order to make further copies of themselves. The actual design of organisms reflects the problems and opportunities that genes face on the path to replication. Biologists use these problems to make predictions about the forms of 'design solutions' or 'adaptations' of which organisms will be composed (Curry, ibid.). It is then possible for biologists to conduct experiments to test for the presence of these adaptations (Dawkins, 1986). Conversely, as Curry (ibid.) suggests, 'biologists can ask whether a particular trait is an adaptation by asking whether it solves (better than chance) a problem that the organism typically faces'. These processes are sometimes referred to as 'engineering' and 'reverse engineering' (Ridley, 1999, 2003). As Tooby (1999) have made clear, 'engineers' begin with a problem and endeavour to design something that will solve the problem. 'Reverse engineers' begin with 'a widget' (Curry, ibid.) and endeavour to work out the nature of the problem. Evolutionary psychology adopts such an adaptationist approach when investigating the design of the human mind. As Curry shows, evolutionary psychologists can begin with a problem 'that would have been recurrent in the lives of our ancestors – such as how to choose fertile mates, or how to maintain co-operative alliances' (ibid.). They are then able to suggest alternative solutions, and design experiments to test for them. As Tooby and Cosmides (1992: 10) suggest, 'If one knows what adaptive functions the human mind was designed to accomplish, one can make many educated guesses about what design features it should have, and can then design experiments to test for them.' For example, using evolutionary theory, comparative data and the ethnographic record, Symons (1979, 1995) was able to make a number of predictions about the evolved design of human sexual psychology. As Curry (ibid.) shows, these predictions were put to the test and largely confirmed by a survey of 10,000 individuals from 37 different cultures conducted by Buss (1999). The process can run in reverse, according to Curry, whereby 'previously mysterious psychological devices can be illuminated by revealing the function that they are designed to perform' (ibid.). An example of this is how Cosmides and Tooby (1992) identified a quirk of human psychology, discovered years earlier on the Watson Selection Task, through a series of experiments as a 'cheater-detection mechanism', a device that game theory predicts is 'necessary for certain forms of co-operation' (Curry, ibid.).

As Curry (ibid.) states, Steven Rose (2000: 247) appears to accept the basic premise of evolutionary psychology:

> The declared aim of evolutionary psychology is to provide explanations for the patterns of human activity and the forms of organisation of human society which take into account the fact that humans are animals and, like all other currently living organisms, are the present-day products of some four billion years of evolution. So far so good.

As Curry (ibid.: 3) shows, Rose (ibid.: 250) continues to expand upon the adaptationist approach:

> Because humans are as subject as any other organism to evolutionary processes, we should therefore expect to find such adaptations among our kind just as much as amongst the others that we study. Individual aspects of being human – from our body shape to our eyes and capacity for binocular vision – are clearly evolved features and fit us to the environment in which we live.

However, as Curry shows, the Roses (ibid.) object to utilising such an adaptationist approach to illuminate the psychological mechanisms that, according to evolutionary psychologist underpin human *social* behaviour. This is because, according to Curry, Rose and Rose claim that 'not enough is known about the conditions under which our ancestors evolved to make claims about the problems that they faced, or to test whether or not particular features of human psychology are adaptations' (Curry, ibid.). In Curry's words, the Roses 'also claim that the period of pre-history focused upon by evolutionary psychologists – the Pleistocene or "Stone Age" – is "the wrong one"' because 'there has been sufficient time since the end of the Pleistocene for significant evolutionary change in the design of the human mind' (Curry, ibid.). In addition, the Roses appear to be arguing that 'evolutionary psychology's claims about universal features of human social psychology are contradicted by cultural and historical variability, and neglect the role of emotion in human mental life' (Curry, ibid.). Finally, it is Curry's view that the Roses mistakenly use

Daly and Wilson's (1998) research on step-parents to represent what they consider to be the empirical deficits of evolutionary psychology.

As we have seen, evolutionary psychologists make and test predictions about human psychological adaptations, based upon information about human evolutionary history. According to Curry, the Roses claim that not enough is known about the Stone Age to render such an approach credible and viable. What do evolutionary psychologists actually know about the Pleistocene, and how do they get access to the information? These are questions posed by Curry (ibid.), and more recently by Owen (2006b). As Curry (ibid.) and Owen (ibid.) have suggested, the work of Tooby and De Vore (1987) shows how information about ancestral adaptive problems comes from a wide range of disciplines which include physical and paleo-anthropology, primatology and other cross-species comparisons, studies of contemporary hunter-gatherer societies, and game-theory models of social interaction. As a result of the work of those such as Cosmides and Tooby (1997), we know that our ancestors nursed, had two sexes, hunted, gathered, chose mates, used tools, had colour vision, bled when wounded, were predated upon, were subject to viral infections, were incapacitated from injuries, had deleterious necessives and so were subject to inbreeding depression if they mated with siblings, fought with each other, cooperated with each other, lived in a biotic environment with predatory cats, venomous snakes and plant toxins. They were omnivorous, ground-living primates and mammals with helpless infants, long periods of bioparental investment in offspring, and an extended period of physiologically obligatory female investment in pregnancy and lactation.

As Curry (ibid.: 3-4) suggests, in order to claim that this body of accumulated knowledge is inadequate as a starting point for evolutionary psychology, 'the Roses would need to show that current research on the Pleistocene is insufficient or unreliable' or the Roses 'would need to argue that the conditions of the Pleistocene are in principle "unknowable", and hence an evolutionary psychology will never be possible'. However, as Curry (ibid.: 4) suggests, 'no such arguments are forthcoming'. According to Curry, the Roses 'do not provide any criticism of any of the actual methods' utlised to investigate the Pleistocene, 'nor do they demonstrate that any of the actual assumptions about ancestral conditions employed by evolutionary psychologists are false, or unreasonable' (Curry, ibid.). As Curry (ibid.) goes on to suggest, the Roses do not 'take issue with any of the predictions or discoveries about adaptations for social life' (the reference is to infanticide, concealed ovulation, sperm-competition, uncertainty of paternity, material-foetal conflict, mate-guarding, 'theory of mind') 'that evolutionary psychologists have made'. In particular, Curry (ibid.) suggests that Steven Rose (2000: 253) 'merely follows Stephen Jay Gould in suggesting that accounts of ancestral conditions are little more than Just-So stories'. For Curry, Hilary Rose's (2000: 118) quip to the ends that evolutionary psychologists offer up a vision of the Stone Age 'that owes more to the Flintstones than to serious scholarship', fails to provide contrary evidence. 'The Roses' contention seems to be merely that finding out what the Pleistocene was like is difficult, and that nothing is certain. But then what branch of science is any different?' (Curry, ibid.: 4). In Curry's view, the Roses are not so much arguing the point as asserting it.

Owen (2006b: 908) shows how Rose and Rose (2000) 'take issue with the idea, expressed by Nesse and Williams (1994), that there is a mismatch between "stone-age minds" and the "space age world"'. Nesse and Williams (ibid.) suggest that human evolution is an extremely slow process, and therefore evolutionary psychologists expect human psychology to be 'designed' to deal with conditions experienced during the million or so years that humans spent as hunter-gatherers, but not with the 'novel' problems encountered during the relatively brief period of roughly 10,000 years that the species has spent as farmers/postoralists, or the even briefer period spent in modern, industrialised, urbanised societies. Nesse and Williams (ibid.) have claimed that the mismatch between 'stone-age minds' and 'space-age world' is part of the reason why human behaviour does not reliably result in increased reproductive success. They cite as an example, our preference for sugars, salts and fats evolved at a time when such valuable nutrients were relatively scarce, and we could not get too much of them. Conversely, these things are abundant in the modern world, and the same properties lead us to suffer from obesity, high, blood pressure and tooth decay. The Rose's (ibid.) second argument takes issue with the assumptions of a 'mismatch', claiming that there *has* been enough time since the Pleistocene for some substantial evolutionary change in the 'design of the human mind'.

Rose (2000: 253-4) suggests that:

> Evolutionary modern humans appeared 100,000 years ago. Allowing 15–20 years as a generation time, there have been some 5,000–6,000 generations between human origins and modern times'. Noting that rapid evolutionary change is commonly observed in non-human species, Rose concludes that the assumption that humans have not changed significantly in the last 10,000 years, 'does not bear serious inspection'.

Firstly as Curry (ibid.: 4) suggests, given that 'the issue at hand is the possibility of evolutionary change during the 10,000 years since the end of the Stone Age, it is not altogether clear why Steven Rose makes an argument about the possibility of change in the last 100,000 years'. For Curry (ibid.: 11), this is evidence of the Roses' 'wider confusion about the dates of the Pleistocene'. As Curry (ibid.: 11) further suggests, '*In Alas, Poor Darwin* they suggest that the Pleistocene was the period between "100–600,000 years ago" (Rose and Rose, 2000: 1). In their *New Scientist* article, the Pleistocene is "between 10,000 and 1.6 million years ago" (Rose and Rose, 22 June 2000). And in their *Guardian* article, they suggest that there has been "100–600,000 years *since* the Pleistocene" (Rose and Rose, 13 July 2000)'.

Secondly, Curry (ibid.: 4–5) suggests that Steven Rose's argument for the speed of evolutionary change rests on 'laboratory and field experiments in species varying from fruit flies to guppies (which) give rates of change of up to 50,000 darwins... English sparrows transported to the south of the USA have lengthened their legs at a rate of around 100,000 darwins, or 5 per cent a century' (Rose, 2000: 253). A 'darwin' refers to changes in the mean of the natural log of morphological characters divided by the elapsed time in terms of millions of years over which the change has occurred (Thain and Hickman, 1995). According to Curry (ibid.: 11), 'at no point does Rose explain how a measure of morphological change can be applied to the kinds of information-processing mechanisms that evolutionary psychologists study'. Curry questions whether Rose is 'perhaps thinking of changes in brain size – but what would it mean for a software package to become 5% longer?' Even if such change had occurred, 'Rose does not explain how or why this would undermine the claim that the mind is adapted to ancestral conditions' (Curry, ibid.). Rose's contention is 'presumably that if a sparrow's leg can become 5% longer over 100 years (approximately 100 generations), then it is reasonable to suppose that humans may have changed (in some unspecified way) by 25% over 500 generations' (Curry, ibid.: 4–5). As Curry (ibid.: 5) makes clear, for this to be possible, the period since the Pleistocene would have 'had to have been characterised by the kind of consistent, uniform, directional selection pressures that led to the changes in the sparrow's legs'. According to Curry (ibid.), Rose 'does not point to any such selection pressures'. On the contrary, Rose (ibid.: 253–4) appears to argue that the period since the Pleistocene has been characterised by 'very rapid changes in human environment, social organisation, technology and mode of production' (Curry, ibid.).

Thirdly, as Curry (ibid.: 5) suggests, quite apart from the question of how quickly evolution could conceivably occur, 'there is the question of whether any such change has *actually* occurred in the human line in the last 10,000 years'. According to Curry, Rose (ibid.) presents no evidence of selection pressures. Rose (ibid.: 253) concludes that 'we really have no idea whether the 6,000 or so generations between early and modern humans is "time enough" for substantial evolutionary change. We don't even know what "substantial" might mean in this context'. As Curry (ibid.) suggests, an obvious test of the Roses' theory would be to look for 'genetic differences in populations that have lived under different conditions since the beginning of the agricultural revolution'. If the Roses are correct, and there has been sufficient time for significant evolutionary changes in the 'design' of human psychology since the Stone Age, then, 'to the extent that different groups occupied different environments, one would expect them to have evolved in different ways' (Curry, ibid.). Curry goes on to suggest that groups of humans that remained in Africa might be expected to differ from those that migrated to the Russian steppes, for example. By implication, hunter-gatherers should posses a different set of 'mental tools' from agriculturalists and industrialists. In the words of Curry (ibid.), 'the Roses do not discuss this question, or present any evidence that bears on its answer'.

Crawford and Anderson (1989) have suggested that one implication of the 'evolutionary' view that the human mind took its current form in the Pleistocene is that all modern humans share a universal human nature. In answer to how evolutionary psychologists reconcile such a claim with 'the manifest diversity of behaviour and culture found around the world', (Curry, ibid.) argues that psychological mechanisms are 'condition dependent', in other words, the behaviour they produce will be different under different conditions. As Curry suggests:

> Suppose for example, that human psychology operates in part according to the rule *'If* resources are scarce, *then* adopt a more aggressive approach to acquiring them'. On this basis one might expect over levels of aggression to vary according to the current economic or ecological circumstances of the people under consideration. Evolutionary psychologists look at, amongst other things, permutations in behaviour in order to work out what the underlying rules are and how they operate. This research, which necessarily involves cross-cultural studies, commits evolutionary psychologists to a strongly 'environmentalist' position: the idea that differences in behaviour are largely the product of differences in environmental-physical, social or cultural factors.

Perhaps here Curry echoes Ridley's (1999: 53) point (in the 'Building Bridges' section of this chapter) to the ends that 'genes need to be switched on', and 'external events – or free-willed behaviour – can switch on genes'. In other words, we should strive to avoid what Owen (2006a, 2006b) and Powell and Owen (2005) call *genetic fatalism* (and which is incorporated into the metatheoretical framework employed here) – the tendency to associate genetic predisposition with inevitability. This point aside, the Roses' third argument against evolutionary perspectives appears to be that cultural and historical variability in cultural forms *refutes* the view from evolutionary psychology that all humans share a universal species-typical psychology. Steven Rose (2000: 262–3), referring to historically-recent changes in female mate-preferences, levels of violence, and fertility in several hunter-gatherer populations, remarks 'Each of these societies has undergone rapid economic, technological and social change in the last decade. What has happened to the evolutionary psychology predictions? Why have these assumed human universals suddenly failed to operate?' However, as Tooby and Cosmides (1992: 46) suggest, 'The recognition that a universal evolved psychology will produce variable manifest behaviour given different environmental conditions exposes this argument as a non-sequitur.' As Curry (ibid.: 6) makes clear, in order to make their point effectively, the Roses 'would need to show not only that behaviour changes, but that behaviour changes in ways that are *not* predicted by the evolutionary psychologists' account of the mechanisms responsible', and 'even if the Roses did present such evidence, they would be refuting a particular hypothesis about a particular mechanism, not the entire evolutionary approach to psychology' (ibid.: 11). It is Curry's view (ibid.: 6) that 'by presenting social and cultural explanations as *alternatives to* – rather than *continuous with* – biological explanations, the Roses perpetuate the myth that genes can operate only in a rigidly "deterministic" fashion'.

Steven Rose's fourth argument appears to be along the lines that evolutionary psychology 'neglects the role that emotion plays in human mental life' (Curry, ibid.). Rose (2000: 262) claims that:

> Emotional mechanisms and indeed their expression are evolved properties, and several neuroscientists have devoted considerable attention to the mechanisms and survival advantages of emotion. So it is therefore all the more surprising to find this conspicuous gap in the concerns of evolutionary psychologists.

As Curry (ibid.: 6) shows, the entry for 'Emotion' in the index of the evolutionary psychologist Steven Pinker's work *How the Mind Works* – to which both the Roses refer – reads as follows:

> Emotion, 65, 143, 315, 363–424; adaptive function, 143, 370–374; facial expressions, 273, 365–366, 374, 379, 414–416, 546; hydraulic model, 57, 65, 551; and imagery, 285; in music, 529, 531–532, 533–534; neuroanatomy, 371–372; phylogeny, 370–371; universality, 364–369. See also Anger, Beauty; Disgust; Exhilaration; Fear; Gratitude; Grief; Guilt; Happiness; Honor; Jealousy; Liking; Love; Passion; Self-control; Sexual desire; Shame; Sympathy; Trust; Vengeance.
>
> (Pinker, 1997: 637)

According to Curry, the Roses also 'neglect to mention' Tooby and Cosmides' discussion of emotion; Buss's work (1994, 2000), Nesse's (1990) paper on the 'evolutionary explanations of emotions';

Ekman's work on the facial expression of emotion, including his recent work involving an edited version of Darwin's *The Expression of the Emotions in Man and Animals* and others. In Curry's words, 'Steven Rose appears to have forgotten that he has debated the evolution of human emotions with Steven Pinker' in Pinker and Rose (21 January 1998). Yet, in Rose and Rose (13 July 2000), the Roses appear to criticise the claim from evolutionary psychology that there is a universality to human emotions. Curry (ibid.: 6–7), argues that Steven Rose's claim that evolutionary psychology 'neglects the emotions' rests on an 'elementary misunderstanding of what it means to be an "information processor"'. For example, Rose (2000: 261) suggests that:

> It is not adequate to reduce the mind/brain to nothing more than a cognitive 'architectural' information-processing machine. Brains/minds do not just deal with information. They deal with living meaning... The key here is emotion, for the key feature which distinguishes brains/minds from computers is their/our capacity to experience emotion.

However, as Curry (ibid.: 7) suggests, the adoption of an 'information processing' or 'computational' approach to the mind 'does not commit you to a desktop-computer view of psychology, any more than charting the parabola of a falling apple commits you to inferring that the apple is performing calculus'. As Churchland and Grush (1999) have made clear, 'Information theory' is merely a branch of mathematics utilised in order to capture how a particular system's inputs map onto its outputs. The theory does not imply that the most accurate explanation of human brain function will actually be in computational terms. As the authors have suggested, in an abstract sense, livers, stomachs and brains *all* 'compute', so consequently 'information' need not exclude 'emotion'. Further evidence (cited in Curry, ibid.: 7) from Tooby and Cosmides (1992: 65) can be provided on this point:

> It is important to keep in mind exactly what we mean by the cognitive or information-processing level... some researchers use it in a narrow sense, to refer to so-called 'higher mental' processes, such as 'emotion' or 'motivation'... In contrast... we use terms

such as cognitive and information-processing to refer to a language or level of analysis that can be used to precisely describe any psychological process: reasoning, emotion, motivation, and motor control can all be described in cognitive terms.

Curry (ibid.) makes the point that *Alas Poor Darwin* does not offer, in the form of research, evidence to 'refute any of the empirical claims of evolutionary psychology', rather 'it carries little sustained discussion of any empirical work'. This is, of course, Curry's own personal opinion. However, Curry does acknowledge that there is an exception here in Hilary Rose's arguments against the work of Daly and Wilson (1998). In the latter work, the authors summarise findings that children are at significantly higher risk of sexual/physical abuse and murder from step-parents than from genetic parents. Daly and Wilson's (ibid.) explication is that parental investment is a costly resource, and that Darwinian natural selection has played a part in the 'design' of parental psychology to preferentially 'invest' in one's genetic children, and to be reluctant to 'invest' in children that are not one's own genetic offspring. As Curry (ibid.) shows, Daly and Wilson (ibid.) make the point that step-relationships can be the site where such reluctance may manifest itself in less than harmonious relations. In extreme cases, such neglect may result in severe abuse or even murder. Although Hilary Rose (2000: 122) does not dispute this higher incidence of abuse and murder by step-parents, her contention is that 'rather obvious matters of context' pertaining to 'psychological strain' and 'financial pressures' of starting second families 'explain better why some men ill-treat their partner's children'. Steven Rose (2000: 260) criticises Daly and Wilson (ibid.) for 'ignoring (these) much more obvious proximal causal processes'. Also writing in *New Scientist* (22 June 2000), the Roses claim that their 'obvious' alternative causes 'are better grounded than untestable evolutionary speculations' (cited in Curry, ibid.: 7–8). Hilary Rose, according to Curry (ibid.: 8), 'does not elaborate on these "obvious" alternatives, nor point to research that has "grounded" them'. According to Curry, Hilary Rose 'does not make any references to any other studies of child abuse whatsoever'. Curry (ibid.) suggests that Rose appears 'entirely unaware that these "obvious" explanations have been put to the test, and that they failed'. Some evidence for this claim may lie in Daly and Wilson's (1998: 29) work on the effects of poverty in which 'this initially plausible hypothesis was rejected, for it turned out that the distribution of family incomes in step-parent homes in the United States was virtually identical to that in two-genetic-parent homes'. Daly and Wilson (ibid.) have also examined the effects of numerous other confounding variables such as 'the age of the child, age of the mother, whether the step-father was present at the birth, and the personalities of people who remarry' (Curry, ibid.: 12). Daly and Wilson (1998: 45–6) also discuss the apparent 'anomaly of adoption'. Daly and Wilson conclude:

> All in all, although several additional risk factors were identified, step-parenthood held its place as the most important predictor (of serious child abuse), and its influence was scarcely diminished when the statistical impacts of all the other risk factors were controlled (p. 31).

As Curry (ibid.: 8) shows, Daly and Wilson (1988) in their major work, *Homicide*, also discuss how 'umbrella' or 'blanket' explanations (overarching theories) such as 'strain' do not explain such nuances as the fact that in families where there are both step-children and genetic-children present, it is the step-children who are invariably singled out for abuse and even murder. According to Curry, Hilary Rose (ibid.) has complained that Daly and Wilson's (2001) data do not distinguish between a step-child whose parent has died and a step-child whose parent is absent because of divorce, or between married step-parents and 'casual lovers'. However, as Curry (ibid.) suggests, Rose 'does not suggest *any* reason why, or any evidence *that*, the distinctions she suggests – or indeed, any of the infinite number of distinctions that one could make – might matter'. Hilary Rose (2000: 121) claims that Daly and Wilson's work receives 'no support from the primatologist Sarah Hrdy, whose work on langur monkey infanticide is key to their thesis'. Hrdy was quoted in *New Scientist* as saying, 'Human violence towards babies and infants may be tragic but it's nothing like what a langur male is doing.' As Curry (ibid.) points out readers of *Cinderella* might be surprised to discover that Daly and Wilson have the same perspective on how Hrdy's thesis relates to their own evolutionary position. As Daly and Wilson (1998: 37–8) have said:

> Human beings are not like langurs or lions. We know that 'sexually selected infanticide' is not a human adaptation because men,

unlike male langurs and lions, do not routinely, efficiently dispose of their predecessors' young... Quite unlike the situation in langurs or lions, human stepfamilies exist in all societies, and most stepchildren survive them.

Curry (ibid.) has suggested that none of the Roses' 'arguments against evolutionary psychology' successfully establishes that the discipline is based upon and relies upon spurious empirical evidence or premises. However, next we need to consider the Roses' accusation that evolutionary psychology relies upon 'unexamined political presuppositions' (Curry, ibid.). Steven Rose (1997) and Rose et al. (1984) appear to promote the view that there is 'more to' evolutionary explications of human behaviour than mere science. Rose et al. (ibid.) offer an arguably Marxist critique of evolutionary biology; and Rose himself (ibid.: ix) warns:

> The rise of the present enthusiasms for biologically determinist accounts of the human condition date back to the 1960s. They were not initiated by any specific advance in biological science, or powerful new theory, but harked back instead to an earlier tradition of eugenic thinking which... had been eclipsed and driven into intellectual and political disrepute in the aftermath of the war against Nazi Germany and its racially inspired holocaust.

According to Curry (ibid.: 9), *Alas Poor Darwin* 'marks a complete reversal from this earlier position', with Hilary Rose now conceding that evolutionary psychology 'eschews any notion of race' and that it is compatible with a reasonably broad range of political viewpoints including 'Peter Singer's Darwinian Left, Matt Ridley's free marketeering, Helena Cronin's feminism, Francis Fukayama's call for state intervention to tackle unemployment'. As Curry (ibid.) makes clear, each of these scholars/researchers provide evidence that facts and values can be kept separate; that one's political goals do not dictate one's science (or vice versa), but, as Trivers (1981) has suggested, once political and social objectives are agreed upon, science can help one achieve these goals. Curry (ibid.) suggests that, in order to demonstrate that evolutionary psychology is politically-motivated, the Roses must prove that 'the political views of a researcher have led to a scientific error', but as yet, in his view 'no such demonstrations are forthcoming' in their work.

Curry (ibid.: 9) makes the powerful point that the Roses fail to show that evolutionary psychology rests upon 'shaky empirical evidence' or 'flawed premises'. Curry's work suggests:

> Evolutionary psychologists use accounts of pre-history corroborated by numerous diverse disciplines; their assumptions about the rate of evolutionary change are in keeping with standard assumptions in evolutionary biology; their claims about human universals are consistent with, and go some way towards explaining, cultural and historical variability; emotions have been the subject of a great deal of evolutionary research; and Martin Daly and Margo Wilson's findings have withstood over two decades of testing and scrutiny. Nor do the Roses provide any reason for thinking that evolutionary psychology is a 'fashionable ideology' motivated by 'unexamined political presuppositions'.

As previously stated, our task here is to establish whether there is sufficient evidence from evolutionary psychology to justify modifying Sibeon's anti-reductionist, anti-relativistic, metatheoretical framework so that it entails focusing upon the *biological variable* (the evidence that human behaviour is, at least in part, genetically determined). The work of Owen (2006a, 2006b, 2007a) has made the case, based in part upon the work of Ridley (1999, 2003), that we can talk in terms of human instincts, and that genes can be switched on, and external events – or free-willed behaviour – can switch on genes. Ridley (2003) makes the case for abandoning the 'nature versus nurture' debate and instead champions the alternative conception of 'nature via nurture'. He believes human behaviour 'has to be explained by both nature and nurture' (ibid.: 3). The author goes on to suggest that the discovery of the human genome has changed everything, 'not by closing the argument or winning the battle for one side or the other, but by enriching it from both ends till they meet in the middle' (ibid.). In other words:

> Genes are designed to take their cues from nurture. To appreciate what has happened, you will have to abandon cherished notions and open your mind. You will have to enter a world where your genes are not puppet masters pulling the strings of your behaviour; a world where instinct is not the opposite of learning,

where environmental influences are sometimes less reversible than
genetic ones, and where nature is designed for nurture.

(Ridley, 2003: 4)

Later in the chapter we incorporate *the biological variable* into the
metatheoretical framework, having considered further evidence from
Ridley (1999, 2003) and others which suggests that human beings can
be simultaneously free-willed (or possessing *agency*) and motivated
by instinct and culture. The chapter, having drawn upon further evidence that nurture depends upon genes too, and how genes need
nurture, plus the relationship between 'free will', genes and 'freedom', concludes with the intention of applying some of the terms
to the study of selected examples of crime and criminal behaviour in
the final chapter. However, at this point we need to re-examine some
of the issues around the *dualism/duality* debate. The emphasis here is
on whether we need a dualistic conception of 'biology' and 'society'
or whether we should adopt an approach which favours a *duality* of
'biology' and 'society' alongside 'agency/structure', etc., in order to
contribute towards a study of crime and criminal behaviour.

Duality versus dualism revisited

As observed previously, Williams (2003: 556) has recently stated that
'material-corporeal' sociologists engaged in the project of 'building
bridges' between sociology and biology appear to be united in the
desire 'to challenge dualism on all fronts'. Benton (2003: 292) has
also called for an approach which would 'dissolve' dualistic oppositions ('nature' and 'society'). One reflects that this is not the same
as Ridley's (1999, 2003) approach which favours a conception of
'nature via nurture'. Ridley argues that human instincts do exist, but
that genes can be 'switched on' by environmental stimuli. Writers of the 'embodied school', such as Newton (2003: 29), appear
to demonstrate both an *oversocialised* approach and *genetic fatalism*
in their criticisms of approaches which acknowledge the existence
of instincts in human beings, and even partial biological causality. Newton's criticisms of Sahlins (1972) and Pollock (1988) centre
around the authors' portrayal of stress as the result of 'natural'
instincts. Newton (ibid.) suggests that 'such arguments present a *crude
dualism* that reduces the "problem" of present day social complexity

to the outmoded biological body', and goes on to argue that 'the biological reductionism of such discourse' is closer to 'reductionist sociobiology than it is to, a putative non-reductionistic "biological sociology"'. Considering the evidence presented so far, and the evidence which is provided later in the chapter from Ridley (ibid.), it is hard to envisage a biological criminology or sociology which denies the existence of instincts. Writers of the 'embodied' school such as Williams (ibid.), Benton (ibid.), Shilling (1993) and Freund (2001) construct *oversocialised* conceptions of 'the body', in the sense that they appear to reject unlearned, genetically based behaviour outright, and tend to view the idea of genetic predisposition as 'inevitability', hence Newton's references to 'biological reductionism' (ibid.). The 'embodied' school appear to argue for a *duality of structure* in which 'biology' and 'society' are seen as 'two sides of the same coin'; for example, Shilling's analysis of the body as *simultaneously* biological and social. Newton (ibid.: 35) sees 'no epistemological reason to erect a barrier between the biological and the social'. Again, this is quite different to Ridley's (ibid.) position because Newton rejects the idea of instincts in human beings, as we may observe in his previously noted criticisms of Sahlins's (ibid.) and Pollock (ibid.). Ridley's approach is to acknowledge the existence of instincts but to acknowledge that genes are 'designed to take their cues from nurture' (2003: 4). Newton's and Shilling's (ibid.) positions appear to be similar to Giddens's (1991a, 1982, 1984, 1991b, 1993) theory of *structuration* in the sense that they are 'elisionist' (Archer, 1995: 93), collapsing 'biology' and 'society' together as Giddensian 'flat' social ontology collapses the 'micro–macro' distinction.

As Sibeon (2004: 48) makes clear, 'the contemporary debates of "dualism versus duality"' centre around critiques of Giddensian structuration theory. Giddens, in company with some poststructuralist and postmodern theorists, considers the micro–macro distinction to be an 'unhelpful dualism' because it serves to polarise social scientists into opposing camps; perpetuates a tendency to associate micro with agency, and macro with constraints, in the form of social structure upon agency (1984: 139, 1993: 7). According to Sibeon (ibid.), Giddens also rejects micro–macro dualism because it tends to 'emphasise the difference between small groups and larger social phenomena'. Giddens appears to favour a *'duality* – not a *dualism* – of action and structure' (ibid.). Structuration theory revolves around the idea of 'duality of structure', in which agency and structure are not separate 'domains'. The distinction between these elements is 'dissolved', as the intention is 'to avoid giving primacy to either action *or* to structure and this is expressed in his concept of "social practices"' (ibid.: 75). With regard to the approaches of Shilling (ibid.) and Newton (ibid.), in relation to their conflation of 'society' and 'biology'; these authors are effectively giving 'primacy' to environmental factors in their denial of instincts. Regarding 'social practices', as Giddens (1984: 2) himself suggests, 'The basic domain of study of the social sciences is neither the experience of the individual actor, nor the existence of any form of social totality, but *social practices* ordered across time and space.' These 'practices' are, as Sibeon (ibid.) suggests, 'a combination of action and structure' and 'structure' is both the medium and the largely unintended outcome of social action/social practices. Perhaps Sibeon (ibid.) is correct when he suggests that 'in social analysis, agency-structure and micro-macro should be employed as *dualisms* that refer to distinct, relatively autonomous phenomena'. Arguably, 'biology' and 'society' should, and can be, employed in analysis as distinct phenomena, although it is important to acknowledge how human social actors can be both free-willed and motivated by instinct *and* culture. To reiterate, it does not seem feasible to deny instinct altogether in the *oversocialised* fashion of Newton's (ibid.) work. As Ridley (2003) suggests, genes predetermine the broad structure of the brain, but also absorb formative experiences, react to social cues and even run memory. Nevertheless, although genes and 'environment' both play a part in the explanation of human behaviour, it may sometimes be necessary to analyse them separately. Adopting a 'duality of structure' viewpoint makes this impossible to achieve. For example, Shilling's (ibid.) analysis of the human body as simultaneously 'biological' and 'social' serves to collapse the terms into an amalgamated 'whole' within which elements cannot be separated. Sibeon (ibid.) makes the cogent point that Giddens 'makes it impossible' to investigate the elements of micro, macro, agency and structure, separately. Perhaps the same can be said about Shilling and other 'material corporeal' sociologists such as Benton (ibid.), Williams (ibid.) and Freund (ibid.). One reflects that it is desirable to be able to investigate biology and the social in terms of their mutual influence upon each other over time, or as Sibeon (ibid.) has said of agency, structure, micro and macro, 'to ascertain the relative impact of each upon any given social situation'. In our case, the acknowledgement of the *biological variable* will play a part in the application of the metatheoretical framework to a study of crime and criminal behaviour.

Margaret Archer (1988, 1995, 1998) plays a central part in the contemporary 'duality versus dualism' debates. She rejects Giddensian concepts of the duality of structure, arguing (as we saw in Chapter 2) for a dualism of structure/action. Here we have a *temporal* separation between structure and action: 'structure precedes the action which reproduces or elaborates it' (Sibeon, ibid.: 102). For Archer (1995: 01), all social analysis and theorising must be based upon dualism as duality is regarded as *'central conflation'*. Benton's (2003: 292) recent attempts to 'dissolve the dualistic opposition between "society" and "nature"' appear to be examples of what Archer regards as *'central conflation'*.

Conversely, Mouzelis has argued (1995: 196) that there *is* room for duality in social theory and analysis, stating that 'I shall continue to use the term "structure" in the way Giddens has defined it'. However, as Sibeon (ibid.) has pointed out, Mouzelis does not so much reject structuration as 'reconstruct it' so that it entails including both dualism *and* duality. Mouzelis has defined duality as a concept that refers to situations 'where actors have no cognitive distance from "rules" (he has in mind situations where there is an unreflective, routine drawing upon rules)', and 'dualism for Mouzelis is a term that indicates, again in a paradigmatic sense, scenarios where actors draw reflexively upon rules for theoretical-analytical, strategic or monitoring purposes' (Sibeon, ibid.: 103). Giddens (1993: 6) maintains that Mouzelis misunderstands his concept of the duality of structure, and that no useful purpose can be served by equating reflexivity with dualism. As Sibeon (ibid.) makes clear, Archer conceptualises dualism as a 'temporal separation of structure and agency, and duality as a collapsing of this separation'. For Mouzelis (and under Mouzelis's metatheoretical framework, Giddens too) the distinction between dualism and duality refers to 'differences in actors' orientations – reflexive or otherwise – to structure' (Sibeon, ibid.). Arguably, the framework employed here (a metatheoretical framework concerned with methodological generalisations as opposed to substantive generalisations) has more in common with the work of Sibeon and Archer in terms of favouring *dualism* and the 'micro/macro', 'biology–society' distinctions than it has with the work of Mouzelis or Giddens. However, as Sibeon (ibid.) points out, Mouzelis's position raises some important points.

Sibeon's interpretation of Mouzelis's position is along the following lines. Mouzelis (1995: 118) claims that Giddensian concepts of 'structure' refer to sets of rules (in a 'virtual' sense) that exist outside time/space. Sibeon goes on to suggest that Mouzelis considers that Giddensian concepts of 'systems' refer to 'syntagmatic, "actual" patterned relations in time-space'. Sibeon's (ibid.) interpretation of Mouzelis suggests that it is 'via the processes of structuration that the paradigmatic becomes syntagmatic'; and, in Mouzelis's (1995: 118) words, 'the term structuration signifies the process by which structures lead to the constitution of social systems'. Mouzelis (ibid.: 120) has suggested that social actors relate to 'structure as "virtual rules"', in terms of duality *and* dualism with 'both equally indispensable'. Here, as Sibeon (ibid.: 104) suggests Mouzelis is not rejecting Giddensian duality of structure, rather he 'extends and reconstructs' the concept. The central point is that Mouzelis sees structuation theory as 'incomplete' (Sibeon, ibid.) as the idea of duality is employed to the exclusion of dualism. Mouzelis's (1991: 100) critique of structuration contains four theoretical terms which are 'continua rather than either–or states of affairs', and they refer to 'variation in the relation of actors to "virtual" structures (as "rules") (the paradigmatic aspect) and, syntagmatically, to "actual" social objects or systems' (Sibeon, ibid.). The concept of paradigmatic duality concerns duality of structure, where social actors are seen as 'drawing on more or less "hidden" rules in a natural-performative manner' (Mouzelis, 1995: 121); and 'it is on this partial case (that is, paradigmatic duality) that Giddens bases his duality-of-structure notion' (ibid.: 121). *Paradigmatic dualism* is, according to Sibeon (ibid.), 'where actors, for theoretical-analytical, strategic, or monitoring purposes distance themselves ("stand back") reflectively from a "virtual" body of rules (that is, in Giddens's and Mouzelis's terms, from structure)'. What is of great importance here is the distance between rules and social actors. *Syntagmatic duality* is 'where the actor is vital to the existence of a social context (that is, to the existence of an "actual" – not virtual – social system, social context or social game (s))' (Sibeon ibid.: 104–5). Here, the social actor plays a part in making/reproducing interaction in a social system. *Syntagmatic dualism* 'refers to states of affairs where the actor is not vital to, or has little effect upon the social content (for example, an office cleaner or junior clerk in a multinational corporation)' (Sibeon, ibid.: 105). As Mouzelis (1991: 100) himself has made clear, syntagmatic dualism can be said to apply in situations where 'actors' orientations focus on interactions or social systems to the production or reproduction of which they contribute but slightly'. Arguably, as Sibeon (ibid.) suggests, despite the 'sophistication' of Mouzelis's typology, the notion of duality as expressed here is actually flawed. It is perhaps more fruitful to endorse Archer's concept of *central conflation*, and employ the micro–macro distinction in sociological analysis. In the case of the synthetic, metatheoretical framework developed here and later applied to the study of selected examples of crime and criminal behaviour, perhaps one should also maintain the 'biology–society' distinction as well as the dualisms of agency– structure and micro–macro.

As we saw in Chapter 2, Derek Layder's (1997) work has contributed greatly both to debates around metatheory in a post-postmodern landscape, and to debates pertaining to the 'duality versus dualism' question. Layder's (ibid.) cogent criticism of Giddensian rejection of the agency–structure distinction suggests that 'Giddens cannot examine how far, in concrete situations, actors are influenced or constrained by social contexts' (Sibeon, ibid.: 106). If one were to employ such a concept of ontological duality in an attempt to 'build bridges' between 'the social' and biology, one would be melding the elements so tightly together that it would be impossible to investigate them separately. Benton's (2003: 293) call for a 'multi-faceted, multi-dimensional viewpoint' which 'avoids the simplistic opposition of "nature" versus "nurture" in relation to health' sounds superficially akin to an ontologically flexible, metatheoretical approach. However, Benton's concept of the 'complex intertwining' of actors' 'bodily contribution, including their genetic endowment, the sociological relations characteristic of their society, and their own position within that society' appears to be a 'sophisticated' case of *conflation*. As Sibeon (ibid.: 113) makes clear, 'Layder's commitment to dualism relates to his contention that social reality is best described in terms of a stratified or "depth" social ontology that involves rejection of any attempts to collapse distinction between micro and macro (or between agency and structure)'.

Layder's social ontology, as we have seen in Chapter 2, includes four 'domains', and each has 'relative autonomy' (Sibeon, ibid.). The terms *psychobiography* and situated *activity* refer to the subjective dimensions of the social context. The terms *social settings* and *contextual resources* refer to the objective dimension. *Psychobiography* is a particularly useful term and concept for those who would make links between biology and the social, as it refers to unique, asocial components of the person. Here, Layder appears to recognise the existence of, at least in part, biological variables and biological causality in human behaviour. Recently, Owen (2012a: 94) has argued that David Garland's (2001) argument for criminal offenders to be treated as 'individuals' would be strengthened by an acknowledgement of 'psychobiography, the largely unique, asocial aspects of an individual's disposition, behaviour and self-identity'. Social *settings* are locales within which situated activity occurs, and the term *contextual resources* describes macro-distributions and ownership of resources, together with widespread discourses and practices. The four 'domains' or layers have 'vertical' or 'lateral' dimensions, and micro, mezo and macro levels of social process are acknowledged. In sum, there are definite links in terms of *dualism* between the framework employed here and the work of Sibeon, Layder and Archer in particular. To reiterate, for the development of the framework it is desirable that we avoid a Giddensian *duality of structure* which would serve to collapse the distinction between agency and structure, micro and macro, and in this context, biology and the social. Rather in social analysis, particularly in the case here of a later application of the framework to the study of crime and criminal behaviour, agency–structure, micro–macro, and biology/'the social' should be employed as *dualisms* that refer to discrete phenomena. This does not mean, of course, that connections cannot be made between variables. In the case of biology and 'the social', it is the case that genes can be 'switched on' by environmental stimuli (Ridley, 1999, 2003).

Incorporating the biological variable

In what follows we examine further evidence pertaining to the relationship between free will (agency), genes and 'freedom' from the work of Ridley and Hamer and Copeland (1999), in particular. The case is made for incorporating the *biological variable*, as a theoretical concept, into the metatheoretical framework alongside *agency–structure, micro–macro* and *time–space*. We have already considered the evidence from evolutionary psychology to the ends that genes *do* play a part in human behaviour, despite the *oversocialised* resistance to the idea from the work of those such as Newton (2003) and Giddens (1993). Here, we also refer back to theorists of the 'embodied' school in relation to their attempts to 'bridge the divide' between the social and the life sciences, and we briefly apply a few terms such as *genetic fatalism* to certain theoretical approaches. It must be emphasised that the detailed application of the framework's constructs to the study of crime and criminal behaviour takes place in Chapter 4. As stated in the 'codification' section in the earlier part of this chapter, the framework is intended to 'prepare the ground' for the study of crime and criminal behaviour, and in order to apply concepts to selected examples, we must first justify its existence, in particular, its modification (from Sibeon's original *anti-reductionist* framework) to include focusing upon genetic or partially genetic (in Ridley's (ibid.) sense of 'nature via nurture') causality.

To recap, the metatheoretical framework employed here focuses upon the original (in Sibeon's work) concepts of *agency–structure, micro–macro, time–space* and Foucauldian *power*, and we are considering whether to incorporate the *biological variable*; the evidence from evolutionary psychology and behavioural genetics for a role for biological causality, at least in part, in explanations of human behaviour. The framework is an attempt at metatheory, which, in the spirit of Sibeon's original *anti-reductionism* seeks to avoid *relativism* and favours methodological generalisations as opposed to substantive generalisations, and *dualism* rather than *duality of structure*. It arises out of a critique of the original forms of illegitimate theoretical reasoning identified by Sibeon; the 'four cardinal sins' of *reductionism, reification, essentialism* and *functional teleology*, alongside 'new' terms such as *genetic fatalism*, or the tendency within mainstream social science to equate genetic predisposition with inevitability and, the *oversocialised gaze*, which refers to harshly 'environmentalist' accounts which deny biological causality, instincts etc altogether. The task of this particular metatheoretical framework is to hopefully generate conceptual devices that will inform the development of substantive theories pertaining to crime and criminal behaviour. Arguably, the 'sensitising device' is nearer in spirit to the work of Dickens (2000) in the sense that it seeks to 'embed' human beings within their 'natural' environment whilst acknowledging both genetic and environmental causality, rather than the work of the 'embodied' school (Benton, Freund, Williams, etc.). The synthetic framework seeks to address the 'reality' that genes *do* influence human behaviour, whilst acknowledging that 'environment' is still extremely important.

As previously stated, the framework arises out of a critique of several illicit forms of theoretical reasoning, one of which is *genetic fatalism*. This tendency to equate biological causality with inevitability is fairly widespread in the contemporary literature pertaining to criminology and sociology and the possibility of a 'marriage' between biology and the social. Newton's (2003: 35) recent, cautious 'call' for a *biological sociology* contains several examples for which evidence is shortly provided. Newton acknowledges Shilling's (1993) 'starting point' for a marriage between the respective fields (an analysis of the human body as simultaneously biological and social), but also illustrates methodological and epistemological problems which hamper any attempts to move beyond Shilling's position. Newton (ibid.: 36) warns against reliance upon 'psycho-physiological research in order to enter the biological terrain', and draws attention to 'difficulties in knowing the body or the biological'. The author makes it clear that he is not suggesting we have arrived at 'some "third way" between critical realism/naturalism and constructionism that can be applied reliably across both the natural and the social arena' (ibid.). The main question for Newton is whether, when crossing the 'divide', we remain on the same ontological and epistemological ground. Newton (ibid.: 37) makes a cogent point when he calls for 'a more sophisticated understanding of psychology and biology' because 'without that, we can neither deconstruct it, nor reconstruct it'. Williams (2003: 559) in reply called Newton's work 'a *welcome, timely* and *topical* contribution' to evolving debates. However, there appears to be evidence for a kind of *genetic fatalism* in Newton's (ibid.: 29) criticism of Benton's (1991: 17) treatment of the work of Walter Cannon (1914). Benton is of the view that Cannon's work reflects a wish to think 'holistically' about life forms. In reply, Newton (ibid.) disagrees, suggesting that Cannon's arguments are based on 'a naturalistic, reductionistic and desocialised account which explains human behaviour chiefly in terms of instincts'. Cannon's (ibid.: 264) suggestion that 'the emotion of fear is associated with the *instinct* for flight, and the emotion of anger or rage with the *instinct* for fighting or attack' is arguably less of a 'reductionistic' and 'desocialised' account than Newton appears to think. Surely, Cannon is arguing for what Ridley (2003: 4) calls 'nature-via-nurture', a scenario in which genes react to social cues, rather than a form of inevitable, fatalistic determinism. Cannon appears to be suggesting that genes 'take their cue' from nurture.

As Ridley (ibid.: 6) has said:

> Genes are not puppet masters, nor blueprints. Nor are they just the carriers of heredity. They are active during life; they switch each other on and off; they respond to the environment. They may direct the construction of the body and brain in the womb, but then they set about dismantling and rebuilding what they have made almost at once – in response to experience. They are both cause and consequence of our actions. Somehow the adherents of the 'nurture' side of the argument have scared themselves silly at the power and inevitability of genes, and missed the greatest lesson of all: the genes are on their side.

Newton (ibid.: 20), when referring to writers of the 'embodied' school (Benton, Freund, Williams, etc.), makes the point that though they eschew reductionist, naturalist and dualist arguments, they 'nevertheless draw on studies that have some or all of these characteristics'. Newton does not appear to appreciate the possibility that one can have a viable framework which emphasises *dualism* (of biology and the social) and acknowledges *the biological variable* (the evidence that some behaviour is, at least, in part, rooted in genetics) without being reductionist. As we have seen earlier, it is possible to acknowledge how genes and environment *both* play a part in explanations of human behaviour, without reducing the explanation to a single, unitary cause, and also maintain a *dualism* (of genes and environment) for purposes of analysis, thus avoiding *conflation*. Newton questions the work of Freund (1998: 854–5; 1990: 464; 1998: 277) and Williams (1998: 129; 2001: 71) for citing work on primates like baboons, in the sense that 'reference to animal studies incites a reductionistic account of human health and appropriates rather inappropriate analytic concepts' (Newton, ibid.: 28). Aside from the *genetic fatalism* implied by the use of the term 'reductionistic' in relation to primate studies,

Newton ignores the 'scientific reality' that Homo Sapiens as a species is an evolved African ape (Dawkins, 1992).

Newton is, arguably, not alone in equating genetic determinism with fatalism in contemporary debates. Williams (1996: 41) criticises medical sociologists for 'their tendency to relegate the body to biology'. Instead, 'sociology should be fundamentally embodied'. Aside from rejecting *dualism*, Williams appears to equate biological determinism with fatalism, hence the use of the phrase 'relegate'. In later work, Williams and Bendelow (1998: 144) make the same mistake of associating any explanation of human behaviour which does not explicitly mention 'environment' with inevitability, when they refer to the work of Freund as 'pointing the way towards' 'a "socialised" biology rather than a reductionist sociobiology'. As one is hopefully making clear, social science appears to be dominated by those who reject biological causality, even in part. As Ridley (1999: 91–2) argues:

> To think otherwise, to believe in innate human behaviour, is to fall into the trap of determinism, and to condemn individual people to a heartless fate written in their genes before they were born. No matter that the social sciences set about reinventing much more alarming forms of determinism to take the place of the genetic form: the parental determinism of Freud; the socio-economic determinism of Marx; the political determinism of Franz Boas and Margaret Mead; the stimulus-response determinism of John Watson and B.F. Skinner; the linguistic determinism of Edward Sapir and Benjamin Wharf. In one of the great diversions of all time, for nearly a century social scientists managed to persuade thinkers of many kinds that biological causality was determinism while environmental causality preserved free will; and that animals had instincts, but human beings did not.

Another illicit form of reasoning which, it is argued here, we should avoid when engaging in analysis is the *oversocialised* gaze. If *genetic fatalism* is the equation of biological determinism with inevitability, *the oversocialised gaze* refers to strongly 'environmentalist' accounts which seek to deny *the biological variable* (biological causality, instincts) altogether. Surprisingly, in the age of the Human Genome Project and rapid advances in the field of genetics, many such accounts may be found within social science. For example, as the work of Owen (2006b: 907) shows, 'Giddens's (1993: 57) suggests that, "human beings have no *instincts* in the sense of complex patterns of unlearned behaviour"'. As Owen (2006a: 190) suggests, 'Foucault's (1980b) argument that sexuality is a socio-cultural creation, that sexuality as we know it is the production of a particular set of historical circumstances and obtains only within the terms of a discourse developed since the seventeenth century, is perhaps another example of the *oversocialised gaze*.' Owen (ibid.) goes on to argue that Foucault's position is very similar to that of the symbolic interactionist writers, Gagnon and Simon (1973). The latter authors adopted a 'radical form of social construction theory which is extremely *oversocialised,* in arguing that there is no "natural" sexual drive in human biological make-up' (Owen, ibid.). Sexual drive must, under their terms, be regarded as a cultural and historical construction. As far as Gagnon and Simon were concerned, not only do we *learn* what 'sex' means, and what is sexually arousing to us, but we also *learn* to want sex. The authors acknowledge that the human body has a repertoire of 'gratifications' (including the capacity to have an orgasm), but this does not mean that we automatically, instinctively want to engage in them. Certain 'gratifications' will be selected as 'sexual' through the *learning* of 'sexual scripts'. From Gagnon and Simon's perspective, socialisation is not concerned with controlling innate sexual desire so that it is expressed in 'civilised', acceptable ways, but the *learning* of complicated sexual scripts which serve to specify circumstances which elicit sexual desire. From this standpoint similar to Foucauldian conceptualisations of sexuality, 'sexual desire' is a learnt social goal. Contradictory evidence, as Owen (ibid.) shows, can be found in the work of Hamer and Copeland (1999: 163) who have shown how genes influence our sexual desire, how often we have sex and 'help make us receptive to the social interactions and signs of mutual attraction that we feel instinctively and now call love'. Importantly, 'the authors, alongside providing cogent evidence for "emotional" and sexual instinctsetc, also provide evidence that genes are not fixed instructions, but rather "take their cue" from nurture/the environment' (Owen, ibid.). Hamer and Copeland (ibid.: 179) investigated whether there was a 'correlation between the D4DR gene and number of sexual partners' in male subjects. The molecular geneticists had previously established that 'novelty seeking' and sexual behaviour are linked, and that 'novelty seeking' is in part mediated by the D4 dopamine receptor gene. They found too that there was indeed a link between D4DR genes and the number of sexual partners in men, 'but in an unexpected way' (ibid.). Looking first at heterosexual men, the geneticists found that the men with 'the long form of the D4DR gene, the high novelty seekers, had slightly more female partners than those with the short form, the low novelty seekers' (ibid.). It should be emphasised that the trait referred to here as 'novelty seeking' means 'finding pleasure in new, varied, and intense experiences' (Hamer and Copeland, ibid.: 178). It was the study by Bogaert and Fisher (1995) at the University of Western Ontario which probably did more than any other towards establishing the idea that a novelty-seeking score was a better predictor of the number of sexual partners than other variables such as masculine age, physical attractiveness and so on. 'The more a person was a thrill seeker, the more partners he had' (Hamer and Copeland, ibid.). Hamer and Copeland's study became rather interesting when they asked how many other men had the respondents slept with. Despite the heterosexual orientation of the respondents, 'some had slept with another man, usually just once and when they were young' (ibid.: 179). Here was a strong correlation to the D4DR gene. 'Straight' men with the long form of the gene, the high novelty seekers, 'were six times more likely to have slept with another man than those with a short gene', and 'about half of the long gene subjects had ever had a male sexual partner compared with only 8 percent of the short gene men' (ibid.). The reverse was true for 'gay' men. As Hamer and Copeland expected, the homosexual respondents 'had more male partners than the straight men did female partners' and 'the D4DR gene had the expected effect' (ibid.: 180). However, the effect of the gene 'was much stronger for the number of female partners of the gay men' (ibid.). Those with the long, high novelty seeking form of the D4DR gene 'had sex with more than five times as many women as did those with the short, low novelty-seeking form'. Although, as the geneticists acknowledge, the gay men *may* have had sexual relations with women in part because of social pressure, 'it seemed that a desire for new experiences also played a role' (ibid.).

According to Hamer and Copeland:

> Those results show that the D4 dopamine receptor gene does influence male sexual behaviour, but indirectly. For a straight man, sleeping with another man is about as novel as you get. For a gay man, having sex with a woman is equally unique. Does this

mean that D4DR is a 'promiscuity gene' and that an errant husband can tell his wife, 'I couldn't help it, it was genetic'? Of course not. A gene doesn't make a person commit adultery. It simply determines the way certain cells respond to dopamine, which in turn influences a person's reaction to novel stimuli. How a person reacts to that stimuli is more a matter of character than of temperament.

(ibid.)

Arguably, Hamer and Copeland (ibid.) provide further evidence that it is a mistake to engage in *genetic fatalism*; predisposition need not imply inevitability. Additionally, it is a mistake to deny genetic influences altogether in the form of instincts in the *oversocialised* manner of Gagnon and Simon (1973) and Giddens (1993). As previously stated in this chapter, we are concerned here, at this point, with establishing whether we are justified in modifying Sibeon's original metatheoretical framework so that it entails acknowledging *biological variables* in any discussion of causality in relation to human behaviour. With this task in mind, we return to the approach of Ridley (1999, 2004), who, it will be recalled, favours an approach to the so-called nature versus nurture conundrum, which can be called '*nature via nurture*', in which genes and environment are accorded *plasticity* and *mutuality*.

In *Genome: The Autobiography of a Species in 23 Chapters*, Ridley (1999) discusses the relationship between genes, environment and freedom in some depth. He maintains that 'the crude distinction between genes as implacable programmers of a Calvinist predestination and the environment as the home of liberal free will is a fallacy' (ibid.: 303). He makes the compelling case that nature can be 'much more malleable than nurture' (ibid.). The author claims that virtually all the evidence purporting to demonstrate how parental influences shape our character is flawed. He acknowledges the correlation between 'abusing children and having been abused as a child', but argues that the link can be accounted for 'by inherited personality traits'. Ridley goes on to argue that the children of abusers inherit their persecutor's characteristics, and correctly controlled for this effect, social studies 'leave no room for nurture determinism at all' (ibid.: 304). As Harris (1998) has shown, it is rare, for example, that the step-children of abusers become abusers. Taking this a stage further, Ridley (ibid.) suggests that this tendency to ignore any possibility of *nurture determinism* characterises 'virtually every standard social nostrum'. Here he refers to the idea that 'criminals rear criminals' and so on.

Interestingly, Harris (ibid.), after a career of writing psychology textbooks, 'suddenly began questioning them a few years ago' to find (according to Ridley, ibid.: 5) that because 'virtually no studies had controlled for hereditability, there was no proof of causation at all in any study'. Correlation was being presented as causation, in other words. Yet as Ridley (ibid.) suggests, from behaviour genetics studies, there was new, strong evidence against what Rich Harries calls 'the nurture assumption'. As Harris (ibid.) has shown in her work, studies of the divorce rate of twins, as an example, reveal that genetics account for around half of the variation in divorce rate, non-shared environmental factors for another half and shared home environment for nothing at all. In other words, as Ridley (ibid.: 305) suggests, 'you are no more likely to divorce if reared in a broken home than the average – unless your biological parents divorced'. Similarly, Harris's (1998) work reveals how studies of criminal records of adoptees in Denmark showed a powerful correlation with the criminal record of the biological parents and a very small correlation with the criminal record of the adopting parents (and that disappeared when controlled for peer-group effects), whereby the adopting parents were found to live in more, or less, criminal neighbourhoods according to whether they themselves were active criminals. Harris has, arguably, played a part in challenging what Ridley (ibid.) calls 'the dogma that has lain, unchallenged, beneath twentieth-century social science: the assumption that parents shape the personality and culture of their children'. As Harris (ibid.) suggests, Freud's psychology, Watson's behaviourism and Mead's anthropology contain assumptions about *nurture determinism* that were never tested. Importantly, as Ridley (ibid.: 307) has made clear, Harris's work 'lays bare just how much more alarming social determinism is than genetic'. Far from 'leaving room for free will, it rather diminishes it', and 'a child who expresses her own (partly genetic) personality in defiance of her parents' or her siblings' pressures is at least obeying endogenous causality, not someone else's' (ibid.).

As we have seen in the section of this chapter titled 'Building Bridges', it is a fallacy to equate genetic determinism and predisposition with fatalism. Yet, 'the myth persists that genetic determinism is a more implacable fate than social determinism' (Ridley, ibid.: 308). As the author makes clear, it makes little sense to argue that 'biological determinism' threatens the case for political freedom. We have seen how Rose and Rose (2000: 107) accuse evolutionary psychologists of being part of a 'restored and re-energised neoliberalism' earlier in the chapter. As Brittan (1998: 10) has argued, 'the opposite of freedom is coercion, not determinism'. Ridley (ibid.) too argues that 'we cherish political freedom because it allows us freedom of personal self-determinism, not the other way around'. For Ridley (ibid.: 313), 'freedom lies in expressing your own determinism, not somebody else's', and it is 'not the determinism that makes a difference, but the ownership'. He goes on to argue that if freedom is what we cherish (and Rose and Rose (2000) appear to attack 'biological determinism' partly because they view it as constraining political freedom), then it is preferable to be 'determined' by forces that originate inside ourselves and not within others. According to Ridley (ibid.), part of the public's revulsion at cloning originates in the idea that what we see as uniquely our own could be shared by another being. Ridley goes on to argue that 'the single-minded obsession of the genes to do the determining in their own body is our strongest bulwark against loss of freedom to external causes' (ibid.: 313). The author makes clear that his optimism lies not in a 'gene for free will' (there is *no* single gene), but rather in the direction of

> [s]omething infinitely more uplifting and magnificent: a whole human nature, flexibly preordained in our chromosomes and idiosyncratic to each of us. Everybody has a unique and different, endogenous nature. A self.
>
> (ibid.)

Ridley makes a compelling case for the *plasticity* and *mutuality* of '*nature via nurture*', genes and environment. We have examined the evidence that suggests sexuality is, at least, partially genetic in terms of causality in the work of Hamer and Copeland (1999). Perhaps, we should also examine, a little more closely, Ridley's (ibid.: 104) claim that '[t]he evidence that grammar is innate is overwhelming and diverse'. Hamer and Copeland (ibid.) suggest that there is growing evidence that genes play a very important role in human language.

One line of evidence comes from 'the fact that language is so specific to our species' (ibid.: 231). Just as human beings possess the greatest cognitive ability of all the species, so they have the greatest ability to communicate thoughts. There is, as the authors suggest, no animal parallel to our complex language that is able to distinguish past, present, and future and express concepts pitched at a very high plane of abstraction. It is not unreasonable then for Hamer and Copeland (ibid.: 232) to suggest that 'there must be something about the human genetic blueprint that makes us especially capable of language'.

As the authors suggest, another line of evidence points to the very complicated rules found in all human languages. All languages have their own precise grammar, syntax and pronunciation etcetera. This is what prompted the linguist Noam Chomsky, according to Hamer and Copeland (ibid.), to theorise that a newborn human baby's brain is 'prewired' with the basic ability to recognise sounds and to learn the rules of grammar and syntax. As the authors show, Chomsky postulated an 'innate language acquisition device'. According to Chomsky's theories, the specific language an infant learns depends completely upon the environment, but 'the basic ability to learn the rules and apply them to the billions of different combinations is genetically preprogrammed' (Hamer and Copeland, ibid.).

Hamer and Copland suggest that one way to 'isolate' the specific genes involved may be to study what happens when the 'language acquisition device' goes 'awry', as in the case of a family sharing a specific language impairment (SLI) which 'mangles' language. In such cases, the authors suggest, children start talking late, and then have great problems articulating words, and make grammatical errors which may persist through adulthood. Such people can have 'normal' intelligence and possess skills in many directions, except language. Specific language impairment, as Hamer and Copeland make clear, runs in families, which suggests the distinct possibility of a genetic basis. The work of Pinker (1994) describes a family in which the grandmother has the impairment, as do four of her five children. They in turn have 23 children, of whom 11 have the 'language problem'. However, as Pinker (ibid.) makes clear, one of the grandmother's daughters speaks 'normally', and so do all their children, as if this 'side of the family' had escaped the gene. According to Hamer and Copeland (ibid.: 233), '[t]his is just the sort of pattern expected for a condition caused by a single dominant gene, although the gene itself has not yet been identified'.

Regarding dyslexia and biological causality, more progress has been made in locating the genetic roots. Once called 'word blindness', dyslexia is characterised by difficulty in learning to read 'despite adequate intelligence and character' (Hamer and Copeland, ibid.). As the authors suggest, it appears to be caused by a 'fundamental brain disconnect between different words and their meaning' (ibid.). The malfunction is specific for reading, and as the authors show, dyslexics can be highly intelligent; they 'just' cannot read adequately. As many as 8% of children are 'reading disabled by standard school system criteria' (ibid.). Children with dyslexia are also the largest group of students who receive 'special' education services, according to the authors. Dyslexia was first recognised as a discrete disorder in 1896, and 'within a decade it was known to run in families' (ibid.). Since then, the authors make clear, family and twin studies have shown 'a substantial genetic effect' (ibid.). Most dyslexics have, at least, one other person in the family who has reading problems, and the rate in identical twins is 40%.

Hamer and Copeland acknowledge that a reading disability is clearly a very complex disorder with many different causes, inherited and environmental. Even so, there appear to be some families in which one major gene is involved. As the authors make clear, scientists at the University of Colorado in Boulder (USA) and at the Boys Town National Research Hospital started a 'gene hunt' in families with at least two dyslexics. They found evidence 'linking reading scores to chromosome 6 and found similar results in fraternal twins' (ibid.). 'More recently, a different group of scientists at Yale University, found linkage to the same chromosome in yet another group of dyslexic families' (ibid.: 233–4). As Hamer and Copeland suggest, these appear to be 'solid results', but it is not yet known which gene is involved or how it works. What does appear to be evident is that:

> The gene has a major effect on the ability to break down long words into syllables, but not on comprehension of short words. That suggests something quite specific – a breakdown in the brain circuit used to segment words – rather than a more general problem. So when the actual gene is found it may tell us something

fundamental about that particular part of intelligence dealing with language and reading.

(Hamer and Copeland, ibid.: 234)

Although Hamer and Copeland claim that the 'gene for grammar' has yet to be isolated, more recent work by Enard et al. (2002) and Lai et al. (2001) suggests that a genetic mutation *may* be responsible for severe language impairment (SLI). As Ridley (2003: 214) comments, 'It is the first candidate for a gene that may improve cultural learning through language.' We have previously seen how SLI has long been known to run in families from the work of Hamer and Copeland. According to Ridley (ibid.), 'there is indeed a gene on chromosome 7, responsible for this disorder in one large pedigree and in another smaller one.' The gene is known as Forkhead box P2, or FOXP2 for short, and according to Lai et al. (ibid.), it is a gene whose job is to 'switch on' other genes – a transcription factor. When it is 'damaged', the person never develops full language.

Dunbar (1996) has suggested that language took over the role that grooming occupies in ape and monkey society – the maintenance and development of social bonds. As Ridley (ibid.) shows, Dunbar (ibid.) argues that when human beings began to live in large groups, it became necessary to 'invent' a form of social grooming that could be done to several people at once: language. Dunbar has noted that human beings do not use language simply to communicate 'useful' information; they utilise it in the main for social gossip. As Ridley (ibid.: 218–9) suggests, 'according to the scanty fossil evidence, speech, unlike manual dexterity, appeared late in human evolution'. As the author shows, the 1.6 million-year-old Nariokotome skeleton discovered in 1984 in Kenya has in the region of the neck vertebrae space for only a narrow spinal cord similar to an ape's, and approximately half the width of a modern human spinal cord. As Walker and Shipman (1996) make clear, modern humans need a broad cord to supply the nerves to the chest for close control of breathing during speech. As Ridley (ibid.: 219) shows, 'Other still later skeletons of *Homo erectus* have high ape-like larynxes that might be incompatible with elaborate speech.' Indeed, some anthropologists such as Tattersall (cited as 'email correspondence' by Ridley (ibid.)) are tempted to infer that, since the attributes of speech appear so late in human evolution, language may have been a 'recent' invention, appearing as little as 70,000 years ago. Arguably, language is not the same thing as speech, and as Ridley (ibid.) observes, 'Syntax, grammar, recursion and inflection may be ancient, but they may have been done with hands, not voice', and 'perhaps the FOXCP2 mutation of less than 200,000 years ago represents not the moment that language itself was invented, but the moment that language could be expressed through the mouth as well as through the hands' (Ridley, ibid.).

It is at this point that we decide whether to modify Sibeon's original *anti-reductionist* framework to include focusing upon the *biological variable*, or evidence from evolutionary psychology and biological science for a, at least in part, genetic basis for human behaviour. After considering the evidence that genes play a role alongside environment, in particular from Ridley (1999, 2003), Hamer and Copeland (1999), Enard et al. (2002), Lai et al. (2001), Dunbar (1996), Pinker (1994), Harris (1998), Bogaert and Fisher (1995), Curry (2003) and Cosmides and Tooby (1997), it is suggested here that there *is* sufficient evidence to warrant incorporating a focus upon the *biological variable* into the 'new' metatheoretical framework, alongside *agency–structure*, *time–space* and notions of Foucauldian *power*. We should keep Ridley's notion of *'nature via nurture'* firmly in mind when focusing upon *biological variables* in analysis, the 'feedback loop' which embraces genes *and* environment, acknowledging *plasticity* and mutuality. We need to remember, drawing upon Ridley (ibid.) and the work of other evolutionary psychologists, that nurture depends on genes, and genes need nurture. Genes predetermine the broad structure of the brain of *Homo sapiens*, but they also absorb formative experience, react to social cues and need to be switched on, and external events – or free-willed behaviour – can switch on genes. We need to avoid *oversocialised* conceptions of the person (Gagnon and Simon, 1973; Giddens, 1993) and the tendency to associate genetic determinism/predisposition with inevitability or *genetic fatalism* (Newton's (2003) treatment of the work of Walter Cannon). We also need to focus upon *dualism* rather than the *duality* favoured by those of the 'embodied school' for example. This is to avoid what Archer (1995: 167–8) calls *'central conflation'*. Sometimes, in the course of analysis, it may become necessary to look at the 'biological' and 'the social' separately, whilst acknowledging links, for example, the point made by Ridley (ibid.) that genes take their 'cue' from nature. Approaches which favour *duality*, such as those of Newton (ibid.), Shilling (1993) and Freund (2001), make the separation of the two variables virtually impossible. This mirrors the *conflation* of *agency* and *structure* to be found in Giddensian *structuration* theory (Archer, ibid.).

Now, in what follows in Chapter 4, we are ready to apply the metatheoretical framework in greater depth to the study of crime and criminal behaviour. All chapters are strongly linked and represent logical stages in the process of developing a sensitising framework and finally demonstrating its explanatory potential. Chapter 4 applies all the meta-concepts in the framework to the study of the area of interest. Examples from criminological and sociological literature are provided, and the task is to demonstrate the usefulness of this post-postmodern, *genetic–social* conceptual toolkit in terms of 'making a contribution towards a return to sociologically based theory and method and suggesting a way forward for criminological theory, and also towards a cautious marriage between the biological and social sciences that is balanced and does adequate justice to the mutuality between genes and environment' (Owen, 2012a: 83).

4
An Application of the Genetic–Social Framework to the Study of Crime and Criminal Behaviour

The metatheoretical framework employed in this book has been previously used to suggest a 'way forward' beyond post-postmodern relativism, in tandem with its application to the study of human biotechnology in the work of Owen (2006b, 2009a) and applied by the author to several other areas of interest such as masculinities, globalisation, ageing, notions of 'trust' and professional power. Here it is applied to the study of crime and criminal behaviour in much greater depth. To recap, the ontologically flexible framework is an example of metatheory. It relies upon methodological generalisations as opposed to substantive generalisations, and multifactorial analysis, preparing the ground for further theoretical and empirical investigation involving large-scale synthesis. The intention now is to show how the framework may be applied and how it may inform criminological theorising, and first we must turn to what criminological theorising must avoid.

The metatheoretical framework

The framework arises out of a critique of the following illegitimate forms of theoretical reasoning:

1. *Reductionism.* This term is included in Sibeon's (1999, 2004) original anti-reductionist framework. Reductionist theories are ones which attempt to reduce 'the complexities of social life to a single,

unifying principle of explanation or analytical prime mover (Hindess, 1986a, 1988) such as "the interests of capitalism", "patriarchy", "rational choice", "the risk society", "trust", "the information society", "globalisation", or whatever' (Sibeon, 2004: 2). An obvious example from Marxist criminological theorising is the reliance upon explanations involving economic determinism.

2. *Essentialism.* This term is included in Sibeon's original framework. *Essentialism* is 'a form of theorising that in aprioristic fashion presupposes a unity or homogeneity of social phenomena' (Sibeon, ibid.: 4). This can include social institutions, or taxonomic collectivities such as 'white men', 'the middle class' or Charles Murray's 'underclass'.

3. *Reification.* This term too is included in Sibeon's original framework. *Reification* is the 'illicit attribution of agency to entities that are not actors or agents' (ibid.). In Sibeon's (ibid.) view, and echoed in the 'new', modified framework, an *actor* is an entity that 'in principle has the means of formulating, taking and acting upon decisions'. Sibeon's original, non-reified definition draws upon Harre's (1981) concept of agency and Hindess's (1988: 45) 'minimal concept of actor', which specify that for an entity to be regarded as an actor, it must be capable of reaching and acting upon decisions. As Sibeon (ibid.) has made very clear, on the basis of such a non-reified definition, there are two types of actors, namely individual human actors and Hindess's concept of 'social actors' (1986a: 115). 'Social actors' include organisations such as government departments, like the Home Office, organised pressure groups and so on, and committees such as the Cabinet and micro-groups such as individual households (Sibeon, ibid.: 5). These 'social actors' have been termed 'supra-individuals' by Harre (1981: 141).

4. *Functional teleology.* Again, this term forms part of Sibeon's original framework. It may be defined as 'an invalid form of analysis involving attempts to explain the causes of social phenomena in terms of their effects, where "effects" refers to outcomes or consequences viewed as performances of "functions"' (Sibeon, ibid.: 6). Sibeon's definition draws upon the work of Betts (1986: 51), and the argument is that if there is no evidence of intentional planning by actors 'somewhere, sometime' (Sibeon, ibid.), then it is a teleological fallacy to engage

in explication of the causes of phenomena *in terms of their effects*. As Sibeon (ibid.: 6–7) suggests:

> All too often, reductionist theorists – including some who subscribe to varieties of 'critical social theory' – begin with a current social or cultural item (a social practice of some kind or, say, a law, a welfare or health system, or a public policy), then attempt to 'work backwards' and claim, without any demonstration of intentional planning by previous actors, that the item came into being 'because', in the view of the theorist, it accorded with the interests of say, a taxonomic collectivity such as 'the upper class' or 'white people' or 'men'.

5. *The oversocialised gaze*. This term forms part of the 'new' modified framework alongside the previous four illicit forms of theoretical reasoning. This 'new' term refers to harshly 'environmentalist' accounts which are characterised by a strong antipathy towards genetic, or partially genetic, explication. The term has been applied to the work of Gagnon and Simon (1973) by Owen (2006a, 2006b, 2007a), in order to criticise the symbolic interactionists' theory that there is no 'natural', sexual drive in human biological make-up. As we have seen in Chapter 3 (the section entitled 'Incorporating the Biological Variable'), contrary evidence can be found in the work of Hamer and Copeland (1999: 163), who have clearly shown how genes influence our sexual desire, how often we have sexual congress and 'help make us receptive to the social interactions and signs of mutual attraction that we feel instinctively and now call love'.

6. *Genetic fatalism*. This is another 'new' term, now incorporated within the framework employed here. As Owen (2006a, 2006b) shows, the term refers to a widespread tendency within social science to *equate genetic predisposition with inevitability*. As we have seen from the work of Ridley (1999, 2003), in Chapter 3 (the section entitled 'Building Bridges'), it is a mistake to view the genes involved in human behaviour as immutable. As Ridley (1999: 153) cogently suggests, 'genes need to be switched on, and external events – or free-willed behaviour – can switch on genes'.

In addition to these 'cardinal sins', the framework focuses upon the *metatheoretical* formulations of *agency–structure, time–space* and *micro–macro*, in addition to notions of Foucauldian *power*, the 'new' term, *the biological variable*, and favours a *dualism* rather than a *duality of structure*. These terms are briefly re-explained as follows. It will be recalled that a *metatheory* is, as Sibeon (2004: 13) suggests,

> intended to inform and hopefully improve the construction of substantive theories and the design of empirical studies. Metatheorists are largely concerned with ontological questions, including the following. What, in general terms, is 'society'? What sorts of things exist in the social world? If there are such things as actors or agents, what sort of things are they? Are activities and society indivisible ('two sides of the same coin'?) and so tightly melded together that (as claimed in doctrines of ontological duality) it is impossible to separate them? Does it make sense to employ a stratified social ontology that refers to micro and macro spheres or 'levels' of society, or is micro-macro – as Foucault, Elias and Giddens claim – a spurious and misleading distinction?

Such questions are the concern of *metatheory*, and of *metatheoretical* concepts such as *agency–structure*. In short, metatheories and meta-concepts are designed to 'equip us with a general sense of the kinds of things that exist in the social world, and with ways of thinking about the question of how we might "know" that world' (Sibeon, ibid.: 13).

7. *Agency–structure*. These are important meta-concepts which refer to significant aspects of social reality. In Sibeon's original framework, and as applied here in the 'new' modified framework, the conception of *agency* is a non-reified one, in which actors or agents are defined as entities that are, in principle, capable of formulating and acting upon decisions. *Structure* refers to the 'social conditions' (Hindess, 1986a: 120–1) or the circumstances in which actors operate, including 'the resources that actors may draw upon' (Sibeon, ibid.: 54). *Structure* then may refer to discourses, institutions, social practices and individual/social actors.

8. *Micro–macro*. This is another meta-concept which 'refers to differences in the units of and scale of analyses concerned with the investigation of varying extensions of time-space' (Sibeon, ibid.).

Micro and *macro* should be viewed as distinct and autonomous levels of social process. Apparently, Sibeon was influenced by Layder's (1994, 1997) argument to the ends that events in social life, at one level, do not determine events at anther level, although 'there may be contingently produced and contingently sustained empirical connections between levels' (Sibeon, ibid.: 55).

9. *Time–space*. This meta-concept refers to 'significant but neglected dimensions of the social' (Sibeon, ibid.: 166). As Sibeon has made clear, the term reflects concerns with temporality and spatiality. Classical social theorists (Durkheim, for example) have tended to regard time as 'social time', distinct from 'a natural essence' (Sibeon, ibid.). However, 'the question of how differing time-frames – including those associated with the macro-social order and those with the micro-social – interweave is a complex matter' that relates to debates pertaining to *dualism* versus *duality* (Sibeon, ibid.). Arguably, as the author also suggests, there is a connection too with Derek Layder's (1994, 1997) (meta)theory of *social domains*, in which the links between objective and subjective domains involve a mingling of 'two different sets of time-frames, namely those associated with (macro) institutional time and those with (micro) situated activity' (Sibeon, ibid.).

The spatial dimension of society here receives attention in relation to notions of indeterminacy and to spatial variation. Sibeon (ibid.: 156) refers to 'the neglect of spatiality' in sociology as opposed to studies of temporality. Soja (1989) too has made this same point. However, as Sibeon (ibid.) makes clear, there has been a certain amount of recent interest in the significance of locale, contingency and spatial variation, notably in community studies (Bulmer, 1985, 1986).

10. *Power*. Like Layder (2004), Sibeon acknowledges the *multiple* nature of power. For Sibeon (ibid.: 145), his (meta)theoretical precepts lead to a view that 'power exists in more than one form; in particular, there are objective structural (including systemic) forms of power, and agentic power', a term used to refer to 'the partly systemic and partly relational and potentially variable capacity of agents to shape events in a preferred direction'. This modified notion of Foucauldian

power favoured by Sibeon has also been applied by Owen (2006a) in the course of his development of a *Post-Foucauldian sociology of aging*, and by Powell and Owen (2005) in their critique of the *Biomedical Model* as employed in social gerontology.

11. *Dualism.* Sibeon's original framework favours *dualism* rather than notions of duality of structure. In doing so, he is also in the company of Layder (1994, 1997, 2004). As Sibeon has suggested (1997a: 72), like Giddensian *structuration* theory, Foucault's work has a tendency to 'compact agency and structure together instead of treating them as a dualism'. Owen (2006a: 186) has employed *dualism* in favour of *duality*, and makes the point that '[t]his Foucauldian tendency (1972, 1980a, 1982) to compact *agency* and *structure* together, to collapse distinctions between the two, results in what Archer (1995) calls "central conflation"'. He (ibid.: 186–7) goes on to argue for the use of dualism in his own work along the following lines:

> The idea of *dualism* as opposed to duality of structure is favoured in the Post-Foucauldian framework, because it is felt that in social analysis, *agency-structure* and *micro-macro* should be employed as dualisms that refer to distinct, relatively autonomous phenomena. For example Shilling's (1993) analysis of the human body as simultaneously 'biological' and 'social' serves to collapse the terms into an amalgamated 'whole' within which elements cannot be separated.

12. *The biological variable.* This 'new' term reflects Owen's (2006a, 2006b, 2007a, 2007b) respectful criticism of Sibeon's original *anti-reductionism* along the lines that it neglects biological or partly biological causality in explanations of social 'reality'. As we have seen in Chapter 3, we should regard the *biological variable* as the evidence from evolutionary psychology and behavioural genetics for a, at least in part, *biological* basis for some human behaviour. We should keep Ridley's (1999, 2003) notion of *'nature via nurture'* firmly in mind when focusing upon *biological variables* in analysis. This refers to the 'feedback loop' which embraces both genes and environment, acknowledging plasticity and mutuality. Genes, as we have seen from the work of Ridley (ibid.) and other evolutionary psychologists in Chapter 3, predetermine the broad structure of the brain of *Homo Sapiens*, but they also absorb formative experiences and react to social

cues. In short, nurture depends upon genes, and genes require nurture. It will be recalled that in Chapter 3 we examined the evidence for a biological basis for certain human behaviour. In particular, the evidence that language is innate and specific to our species was examined with reference to the work of Hamer and Copeland (1999), Ridley (2004), Dunbar (1996) and others, and the work of Hamer and Copeland (ibid.), and Bogaert and Fisher (1995) in which evidence that sexual drive is genetically determined as opposed to being a cultural and historical construction (Foucault, 1980b; Gagnon and Simon, 1973) was also examined. Combined with additional evidence from the work of those such as Harris (1998) which challenges *nurture determinism* with hard evidence from studies in the field of behavioural genetics upon personality traits, the decision was made to include *the biological variable* – evidence for genetic or partly genetic causality as a concept in the 'new' modified metatheoretical framework.

Now, the task is to demonstrate the framework's explanatory potential by identifying examples of the illegitimate forms of reasoning listed previously within the literature and approaches to criminological theorising within contemporary social science and to examples of related bio-social issues such as the 'bridge building' between social and biological sciences. This will also involve an application of the metatheoretical concepts to the literature and approaches. Selected examples from criminological theory are examined, and prominent 'schools' of social science which attempt to 'build bridges' between the social and life sciences such as *Actor-Network theory* and *Posthuman* approaches are also looked at through the lens of the 'new', metatheoretical framework. As Delanty (2002: 1) has cogently suggested, the social sciences 'cannot afford to ignore the challenge that is coming from the new genetics', and 'this will involve rethinking some of the epistemological assumptions of modernist social science'. Delanty (ibid.) has usefully pointed to 'a possible opening within the sciences for a reorientation for sociology and in a way that the new genetics can be addressed in a serious way'. This lies in 'a qualified version of constructivism' (ibid.). The argument put forward in this book, supported by the work of Owen (2006a, 2006b, 2007a, 2007b, 2009a, 2012a) and Powell and Owen (2005), is that another 'opening' may lie in a 'bridge building' exercise in which we acknowledge the *mutuality* and *plasticity* of genes and environment. This, of course, involves acknowledging *the biological variable,* or the evidence from evolutionary psychology/behavioural genetics demonstrated in Chapter 3, for biological causality. Yet, predisposition, as Ridley (1999, 2003) cogently suggests, need not be viewed through the lens of *genetic fatalism* – genes respond to social cues and are not immutable. Arguably, a post-postmodern, metatheoretical framework which employs Ridley's (ibid.) *'nature via nurture'* approach is capable of contributing both to the 'return to' sociological theory and method associated with Layder (1994, 1997, 2007), Sibeon (1999, 2004, 2007) and others, and also to the study of crime and criminal behaviour. The 'pay off' will hopefully lie in the framework's explanatory potential in terms of further theoretical development.

Reification

Reification, the 'cardinal sin' of illegitimately attributing *agency* to entities which are not actors, can be encountered within the literature of contemporary criminological theorising. An actor is an entity that in principle 'has the cognitive means of formulating, taking and acting upon decisions' (Owen, 2012a: 91). Therefore, 'the state', 'the Metropolitan Police', 'society' and so on are not regarded as actors. This approach, particularly with regard to conceptualisations of the state as a non-actor, places the framework very much at odds with conceptions favoured within much of left idealism.

At this point, it is necessary to consider the important matter of *non-agency*. We need to keep this firmly in mind as in what follows we examine the 'schools' of *Actor-Network theory* and *Posthumanism*, which both have conceptions of the actor which are greatly at odds with the *non-reified* conception favoured both by Sibeon (2004) and in the 'new' modified framework. To reiterate, we are applying Sibeon's original conception of an actor as an entity that, in principle, has the means of formulating and acting upon decisions to the study of crime and criminal behaviour. 'Schools' such as *Actor-Network theory* and *Posthumanism* both, arguably, attempt to redefine the relationship between the social and nature and are regarded as related to sociologies of science. It is argued here that their *reified* conceptions of the actor make the work of the schools less useful as conceptual frameworks with which to study crime and criminal behaviour or to 'build bridges' between the social and life sciences. In relation to the 'cardinal sin' of *reification*, Sibeon (ibid.: 122) argues that 'agency is not synonymous with social effects'. There are lots of phenomena possessing social effects (the author provides the example of a conditioning effect upon agency), but they are not necessarily actors. Sibeon (ibid.) makes the point that many different forms of phenomena, such as 'magazines, rain, the HIV virus, the River Nile, motor cars, the Internet, and the Moon', have 'social consequences', but to label them actors equates to the illegitimate attribution of agency. He (ibid.) identifies *reification* in the work of Law (1991a) pertaining to his conception of agency. Law's work is seen as closely influenced by *Actor-Network theory* here, alongside structuralism, poststructuralism and so on. According to Law (1991b: 173–4), 'an agent is a structured set of relations with a series of (power) effects'. As Sibeon (ibid.) points out, this does not include strategies and intentions, and he asks how '"social relations" formulate, take and act upon decisions'. Sibeon makes it clear that he cannot agree with either Callon's use of poststructuralism and Actor-Network theory in regard to Callon's (1986: 204) claim that scallops are actors, or Callon's (1991: 142) claim that the Chernobyl nuclear plant is an actor. As Sibeon (ibid.) suggests, such *reified* claims can be found in some of the more mainstream areas of social science literature. For example, he cites the work of Clegg (1989: 200) who claims that computer systems display agency, and that 'agency may be vested in non-human entities as diverse as machines, germs... and natural disasters'. As Sibeon (ibid.) says, such views are misplaced and should be rejected.

Recently, there have been attempts to extend *Actor-Network theory* beyond studies of technology, power and organisation of the body into dialogue with feminisms, anthropology and psychology by Law (1999). Williams-Jones and Graham (2003) have claimed that social, ethical and policy analysis of the issues arising from gene patenting and commercial genetic testing using human biotechnology is enhanced by the application of *Actor-Network theory*. They appear to be very much influenced by the approaches of Callon and Latour (1981), Callon (1986) and Latour (1987). As Bucchi (2004) has said of this 'school', for the proponents, science has two faces, 'like the Janus of Roman mythology: on the one hand there is "ready made" science; on the other, science "in the making" or research', and whilst it is 'the task of epistemology to analyse the characteristics of the former', it is 'the task of the sociology of science to study the latter' (ibid.: 70). Williams-Jones and Graham, in their recent (2003) application of *Actor-Network theory* to the ethical analysis of commercial gene-testing, appear to be particularly influenced by the distinction (perhaps seen to best effect in Latour (1987)) between human and non-human actors. As Bucchi (ibid.: 72) suggests, for the proponents of Actor-Network theory:

> A research colleague, a bibliographical citation in a paper, an apparatus which yields a microscope image, a company willing to invest in a research project, a virus that behaves in a certain way, the potential users of a technological innovation: all these are allies in the process that transforms a set of experimental results and statements or a technological prototype into a 'black box': a scientific fact or a technological product.

Bucchi (ibid.: 70) also suggests that *Actor-Network theory* can be viewed 'as an attempt to expand the explanatory capacity of the microsociological approaches to science'. Williams-Jones and Graham (ibid.) take the theories and go on to suggest that there is potential in transferring *Actor-Network theory's* 'flexible' nature (ibid.: abstract) to an applied heuristic methodology for 'gathering empirical information and for analysing the complex networks involved in the development of genetic techniques'. The authors explore three concepts in their *New Genetics and Society* paper – Actor-Network Theory, translation and drift – and apply these to the case of Myriad Genetics and their commercial BRACA analysis genetic susceptibility test for hereditary breast cancer. If we apply the meta-concept of *reification* to their work, we can clearly observe that the authors commit the 'cardinal sin' of reification – the illegitimate attribution of *agency* to entities which are not actors. Williams-Jones and Graham (ibid.) treat the susceptibility test as an active participant in socio-technical networks. This is rather like Callon's (1986) treatment of scallops. The authors suggest (ibid.) that the test 'interacts with, shapes and is shaped by people, other technologies, and institutions', and that 'such an understanding enables more sophisticated and nuanced technology assessment, academic analysis, as well as public debate about the social, ethical and policy implications of the commercialisation of new genetic technologies'. It is difficult to see how the authors could possibly justify their claim, because it is surely based upon a *reified* notion of *agency* on the part of 'the test'. How a test for the susceptibility of breast cancer can be capable of formulating and acting upon decisions is not explained by Williams-Jones and Graham (ibid.).

Recently, Pickering's (1995a, 1995b) *Posthumanism* has suggested some original conceptual resources for theorists. As Jones (1996: abstract) suggests, Pickering 'tenders nothing less than a fundamental reordering of social thought', by offering a 'decentred variant of constructivist sociology of science', which invokes the concept of 'material agency', and seeks to 'redefine the relationship between "Nature" and "Society", whilst dismissing the "humanist bias" inherent in sociological inquiry'. As Sibeon (2004: 122) has made clear, he repudiates Pickering's (2001) notion of *posthuman agency*, 'which attributes agency to machines and physical objects', as *reification*. Pickering, as Jones (ibid.: 291) points out, allies himself 'in qualified fashion' with the *Actor-Network* programme in science studies championed by those such as Callon (1986) and Latour (1988).

As Jones (ibid.) argues, *Actor-Network theory* 'finds its conceptual footing in the semiotics of French structuralist and poststructuralist thought'. It is, in essence, a form of 'antihumanism', depicting sociological explication/description as 'illegitimate impositions of modernist categories' (ibid.). Callon and Latour (1981) contend that moderns having been subject to scientific representations must abandon human-centred forms of inquiry/expression if they are to develop what Jones (ibid.) terms 'genuinely holistic appreciations of social life in its material settings'. Callon and Latour (1981) suggest instead a method of theorising associations between people and things that eschews all established ways of speaking about such relations. A decentred 'technoscience' is given primacy, in which 'society is no longer distinguished from the scientific, the technological, or the natural' (Jones, ibid.). The object of analysis becomes a 'unified field of "heterogeneous" networks that stitch together elements drawn from each of these "counterfeit" categories' (ibid.). By 'disrupting commonsense appreciations of "mankind", "nature" and "society" – indeed by rejecting such classifications altogether', Callon and Latour (ibid.) seek a 'new' approach to investigating the ways in which social life is 'enmeshed in technological projects' (Jones: ibid.). Jones makes the point that Pickering 'concurs, generally, on the need for this strategy' and notes that Pickering (1986: 561) has suggested that 'sociology has traditionally focussed upon human individuals and groups as the locus of understanding and explanation', and to remedy the problem as he sees it, now recommends a *'posthumanist displacement of our interpretive frameworks'*.

Jones (ibid.: 292) makes the point that it is the concept of 'non-human agency', the idea that natural objects 'act' as human beings are able to do, that is 'the linchpin in the antihumanist argument'. For Jones, Pickering's point of view appears to be that neither sociology nor poststructuralism, nor *Actor Network theory* 'can relate or distinguish with complete adequacy the "performative" (ie concrete and temporal) qualities of things material, human and nonhuman' (Pickering, 1995b: 9–15). Jones (ibid.) poses the question, 'Has he got this right?' For Jones, the answer is in the negative. As far as we are concerned, in relation to the meta-concepts of the framework, the answer lies in the negative too. We employ here, a tightly-drawn, *non-reified* concept of the actor that borrows from Sibeon (2004). Sibeon (ibid.), it should be noted, acknowledges social actors (organisational actors) in addition to individual actors, who, in the latter case, must be able to (in principle) formulate and act upon decisions. Pickering (1995b) is arguing for a *material agency*–material relations between people and things can be considered in terms of *human agency*. Sibeon (ibid.: 147) usefully discusses the issue of whether machines are actors, in his case, specifically computer systems. The author makes the point that 'inside many aircraft there are "automatic pilots" and in all major airports air traffic control systems are computerised'. He goes on to mention that in the field of medicine 'software is available for assisting in the diagnosis of illness', and that the 'industrial application of robotics, in car manufacturing, for example, is highly advanced'. However, as Sibeon importantly makes clear, 'in each of these cases' the computerised decision-making is wholly 'dependent on hardware that is made and maintained by humans'. The computer software is 'bounded' by parameters that are determined and monitored by human programmers (ibid.). Sibeon does not deny the possibility that 'advances in computing hardware, software and bio-engineering might in future prompt some revision of the anti-reified conception of actor (agent)' (ibid.). However, at this stage, it is merely *reification* to concur with Pickering's conception of *material* or *non-human agency*. Similarly, it is the contention here that it would be a mistake to concur with Brown's (2009: 488) claim in relation to cyber-crime and 'Virtual Criminology' to the ends that critics of the latter school 'play down or ignore the merging of the human and the technical through sociotechnical environments such as the dissolution of the body into information, disembodied identities, digitalizing the human, simulated consciousness, and cybernetics'. Arguably, it is not so much that the critics of Virtual Criminology such as Grabosky (2001) downplay or ignore such things, but that they appear to realise that the agency in all cases comes from the human element. Computers are not actors capable of formulating and acting upon decisions and require programming by human agents.

Jones (ibid.: 293) emphasises that *Posthumanism* 'refers to deliberate rejections of "scientific realism" (the belief that nature alone determines the ways in which scientific representations are developed and refined), *and* "social realism" (the idea that questions of scientific fact are never answered definitively by nature but are decided instead by human interactions)'. As the author suggests, in this sense, it resembles poststructuralist attempts to 'decompose and negate the epistemic warrants of all natural and social scientific inquiries' (ibid.). Poststructuralism, as Jones suggests, does not provide any resources for 'conceptualising material phenomena as they take on form and substance in "real time"'. In their rejection of both social and natural sciences, they provide no substitutes. Here, as Jones (ibid.) suggests, *Actor-Network theory* differs from poststructuralism. 'They *do* propose a materialist alternative to scientific (and commonsense) descriptions', claims Jones. For the author (ibid.: 294):

> Latour suggests that modern science, interpreting the world as it does in Cartesian and Kantian terms, paints an inadequate portrait of human and nonhuman *realities* (1993). Through the opposition of subject and object, society and nature, science misrepresents the world. Its categories are carved from reality in objectionable ways: science fails to capture the essential continuity or 'symmetry', of people and things. As a remedy, actor-network theory recommends that analysts grant to material objects the semiotic equivalent of human agency. But the kinds of semiotics that Latour proposes differs from the poststructuralist variety in important ways (see Latour 1993: 63–64). Actor-Network theory asserts

that signs must be interpreted not only in relational terms – that is, within the logic of a closed semiotic system – but now also in terms of empirical conditions.

Pickering appears to borrow from the 'decentred perspective' of the *Actor-Network* scheme, but as Jones (ibid.: 295) points out, he 'cannot take seriously notions of metaphysical or nonhuman intentionality'. On one hand, as Jones (ibid.) shows, Pickering (1993: 565) has claimed that 'we humans differ from nonhumans precisely in that our actions have intentions behind them'. Yet, on the other, he still attempts to 'preserve a decentred outlook' (Jones, ibid.) when he recommends that we 'acknowledge a role for nonhuman – or material, as I will say – agency in science' (1993: 562). Pickering, as Jones (ibid.) cogently suggests, 'is searching for a way to displace the human agent without envisioning a semiotic subject that vanishes entirely from the field of free action'. In other words, under the 'new' framework's definition of actor, Pickering is still engaging in illegitimate theoretical reasoning – *reification*.

Jones (ibid.: 297) makes what many might consider a cogent point when he suggests that 'it is a mistake to spend time puzzling over the meaning of nonhuman agency in posthumanism or actor-network theory – the important question is whether individuals are granted some measure of autonomy, or alternatively, are represented as subjects of structural logics'. This point is important because 'the preservation of asymmetric human agency' is the real purpose of posthumanism, and the reason Pickering distinguishes his approach from that of Callon and Latour's (1981) *Actor-Network theory*. As Jones (ibid.: 297–8) suggests, 'the conceptual viability of material resistance depends on the retention of purposeful human activity'. In *posthumanism,* resistance tends to 'emerge' only *as* resistances *to* autonomous human agents (the tendency here is still firmly towards *reification*). As Jones (ibid.: 298) argues, resistances cannot be reconciled with antihumanism, and 'this is the structure/agency problem all over again', arising from the *Posthumanist* tendency to draw 'conceptual resources from discordant theoretical tradition'. For Jones, Pickering's project might be better served by 'reconsidering the enduring questions in the human search for self-knowledge', because 'posthumanism is a form of forgetfulness' (ibid.: 306). Jones (ibid.) explains himself thus; 'by surveying established traditions of social thought, unproductive detours through "new improved" brands of inquiry can be avoided'. Jones refers to recent attempts to suggest that the foundations and purposes of sociological theory and method are 'suspect' by poststructuralists and postmodernists, and entreats us to 'accept no substitutes' for the discipline and methods of sociological tradition. One certainly cannot accept the *reified* conception of the actor employed by *Posthumanists* and *Actor-Network theorists*, and the framework employed here is intended to be a contribution towards the post-postmodern 'return to' sociological theory and method. However, that does not mean a return to 'sociological business as usual'. It must be emphasised here that a metatheoretical, flexible ontology is favoured, and an approach which combines 'sociological realism' with an acknowledgement that genes *do* influence behaviour. This is not the same as a recommendation that we return to the *oversocialised* conceptions favoured in some strands of 'mainstream' sociology and criminology. An example here would be Giddens's (1993) apparent denial of human instincts in the sense of unlearned patterns of behaviour.

To recap, this approach to reification places the framework at odds with conceptions of agency favoured within much of left idealism. It is also at odds with Garland's (2001) over-reliance upon a reified notion of agency (Owen, 2007a, 2012a). Like Foucault, Garland appears to presume that discourses are themselves social actors in the sense of a 'crime consciousness' diffused through the media. Arguably, however, discourses are a form of material that needs to be manipulated by social actors. In *genetic–social* theorising, we may conceptualise discourse as a potential influence upon social actors, but to regard discourse as an agent is to engage in illicit reification. When Garland (2001:120) suggests that the punishment of criminals is 'the business of the state', which he regards as symbolic of 'state power', and that 'the state' is responsible for the care of offenders, he is engaging in reification. As has been hopefully made clear here, the state cannot be regarded as an actor in the sense that Garland appears to imply, although its functionaries may be individual social actors.

In what follows, we apply the meta-concept/ 'cardinal sin' of *essentialism* to the study of crime and criminal behaviour. Our task here, to recap, is to demonstrate the conceptual usefulness of the 'new' modified framework.

Essentialism

Essentialism as employed in Sibeon's (2004: 4) system refers to 'a form of theorising that in aprioristic fashion presupposes a unity or homogeneity of social phenomena, such as the law or some other social institution' or taxonomic collectivities such as 'women' or 'white people', Murray's 'underclass' and so on. The term *essentialism* is used here to refer to the illegitimate attribution of homogeneity to social phenomena on a priori grounds. This is *distinct* from 'the uses of this term which are to do with the question of whether phenomena – social categories such as "women" for example – have real essences or are socially – constructed' (ibid.: 197).

It could be argued that some of Foucault's ideas have essentialist undertones. For example, in his 'structuralist' period that Foucault (1970) does not adopt Levi-Strauss' concept of universal unconscious structures. Instead, Foucault suggests that unconscious codes/rules underpin specific epistemes and associated discourses. Epistemes are not seen as universal and absolute, but rather as characteristic of particular societies. As Sibeon (2004: 71) has argued, 'usually there exist numerous conflicting and interacting politics of truth within a society'. Padgett and Ansell (1989: 33) make the point that a problem with Foucault's overly-holistic theory is that it fails to address and acknowledge the discontinuities between macro-cognitive and micro-behavioural spheres of the social world, which results in a perspective which 'rips individuals out of their (often contradictory) multiple network contexts and obscures...heterogeneity and complexity'. Perhaps the previous criticism lacks impact when applied to Foucault's (1980b) 'genealogical thesis', which argues against linear, evolutionary historical progression from one episteme to the next. Therefore, as there is no immanent direction to history, meanings associated with a particular era are contingent outcomes of events and struggles. O'Mally et al. have usefully drawn attention to the ambiguities inherent in Foucault's (1991) work, suggesting that elements of Foucauldian analysis veer towards holism and essentialism, whilst his critique of modernist theorising implies heterogeneity/diversity.

Essentialist reasoning can be encountered in the work of several radical feminist criminologists (Dobash and Dobash, 1992; Hester et al., 1996), in the sense that such theorists do not appear to regard the degree of homogeneity or heterogeneity of social phenomena as an empirical variable for investigation, but rather presuppose on theoretical or political grounds a necessary unitariness of the phenomena under investigation, in this case patriarchy. This is clearly linked to a corollary of reductionsm, and similar to reductionism, essentialism may take on a 'disguised' form as, for example, when men are described as social categories that, though perhaps acknowledged to be in some respects internally divided, should be regarded as ultimately possessing an overriding commonality in the sense of patriarchy that transcends all other memberships or affiliations. As we have observed previously, unitariness should be regarded as a contingent, emergent and possibly temporary outcome of social processes rather than a necessary effect of social totality in criminological theorising.

Sibeon (1996: 34) makes the point that although *essentialism* is clearly related to *reductionism* it has a tendency to be more specifically linked 'to the reductionist notion that taxonomic collectivities – such as "women" – are a relatively homogeneous category comprised of individuals (individual women, in this instance) who have more or less common (and "objective") interests' that are 'given' to them because of their structural position in the taxonomic collectivity. The author suggests that *essentialism* can be found in abundance in contemporary feminisms, for example (ibid.). As he goes on to show, *essentialist* feminist theories of which Elliot and Mandell's (1995) work is cited as an example, mistakenly suppose that '"women" *as represented in the theory* is a social category that is empirically "real"', in other words a category given by biology/society rather than the 'product' of social construction or a feminist theoretical construct (ibid.). With these points in mind, it is therefore possible to identify the 'cardinal sins' of *essentialism* within certain feminist critiques.Interestingly, Hanmer (1993) refers to how the Western 'birth control movement' was influenced by Galtonian eugenic ideas and their political expression, in addition to how the movement may be perceived as a grass-roots movement to enable women to space their children in order to protect their health. In doing so, she arguably engages in *essentialism* and *reification* when she refers to 'this double edge of control by women and of women by the state remains an unresolved issue today' (ibid.: 231). Hanmer refers to women as if they are a homogeneous category whilst attributing *agency* to 'the state', which is an entity most definitely *not* an actor according to the definition employed by Sibeon, and in the 'new' framework outlined here. She also draws attention to a critique common to feminisms, in which the notion of choice for women is challenged by 'arguing that the pressures on women to have children, but only when it is socially acceptable to do so' (ibid.: 232) are so great that it is pointless to discuss issues of 'choice'. Again, the emphasis is upon women as an *essentialist* monolithic block. 'Women are seen as having no option about whether or not to have babies', and 'motherhood is socially compulsory, even in countries with coercive population control policies' are phrases (ibid.) which further indicate Hanmer's *essentialist* tendencies.

In what follows, we apply the meta-concept of *Functional Teleology* to the study of crime and criminal behaviour. This is, as was the case with *reification* and *essentialism*, to demonstrate the framework's explanatory potential and the usefulness of specific meta-concepts in relation to criminological theorising.

Functional teleology

In Sibeon's (2004: 6) original scheme, and as employed here in the context of the 'new' metatheoretical framework, *functional teleology* is 'an invalid form of analysis involving attempts to explain the causes of social phenomena in terms of their effects'. This is where 'effects' refer to *outcomes* and *consequences* 'viewed as performances of "functions"' (ibid.). Sibeon appears to have been influenced by Betts (1986: 51), whose work has suggested that it is illegitimate to try to explain the causes of phenomena 'in terms of their effects', in the absence 'of a demonstration of intentional and successful planning by actors' (Sibeon, ibid.). Sibeon (ibid.) cautions against the *reductionist* tendency (particularly prevalent amongst those who 'subscribe to varieties of "critical social theory"') to attempt to 'work backwards'; that is to say, to claim without evidence of intentional planning by previous agents that a certain phenomena came 'into being' because it accorded with 'the interests of, say, a taxonomic collectivity' such as 'the middle class' or 'white men' and so on. As the author cogently suggests, *teleology* is flawed 'by a problem of logic' in the sense that the causes that 'create' a social/cultural phenomenon must *predate* the existence of the phenomenon (ibid.). Consequences or effects can surely only occur in respect of phenomena already in existence.

With this concept of *functional teleology* in mind, we can 'apply' it to the study of, and literature pertaining to, crime and criminal behaviour. As Sibeon (1996: 20) has suggested, 'Marxism and radical feminism, have been bedevilled by the problem of teleological explanation.' One controversial Marxist example is Hall et al.'s (1978) *Policing the Crisis*, which attempted to 'rework many of the utopian aspects of *The New Criminology* into a more modern and sophisticated theoretical package' (Hopkins Burke, 2009: 189). This approach reflects transitions within Marxist theory during the 1970s which encompassed culture, ideology and hegemony into a more complex analysis of crime and criminal behaviour. Hall et al. (ibid.) incorporated ideas from the Italian Marxist Antonio Gramsci in a discussion and analysis of the issue of street robbery or 'mugging'. In the course of this, Hall's Birmingham 'School' investigated the relationship between ethnicity, social class and the state, focusing on the claim that 'mugging', and specifically the stereotype of the young, black man as potential 'mugger', 'marked a social crisis, a breakdown in social order, that could only be resolved by the tougher policing and sentencing of the young men who were its perpetrators' (Clarke, 2013: 30). As Clarke (ibid.) suggests, Hall and colleagues moved outwards from one specific case in the 1970s to examine the different contexts that came together to encourage the development of a so-called 'Law and Order Society'. These included the mass media; the role of political and juridical actors in the social construction of the 'black mugger' as a folk devil and national menace; the role of imagery of street crime; and the UK's multiplying economic, social and political 'crises' as it moved towards a 'crisis of consent', interpreted by Hall et al. as a 'crisis in hegemony' in which 'the imagery of the British social order (and its descent into disorder) was profoundly racialized' (Clarke, ibid.). Unfortunately, it appears to be the case that Hall et al. (1978) are engaging in *functional teleology* here. Their account of authoritarian populism and the use of the stereotype of 'the black mugger' as a hegemonic device does not provide evidence of intentional planning by agents 'somewhere, sometime', and therefore must be regarded as an interesting but essentially flawed study. To reiterate, it is a teleological fallacy to attempt to explain the causes of phenomena in terms of their effects in the absence of such evidence. There are links between the approach favoured here and the Realist ontology of Alain Robbe-Grillet (1963) in the Heideggerian sense of a 'theory of pure surface' positioned against the exaggerated emotionalism of some meta-narratives and the symbolic thinking that characterises theories of hegemony favoured by some Marxist criminologists.

Foucault (1972) strays into the territory of illicit *functional teleology* particularly with regard to the concept of 'disciplinary technologies'. Here Foucault asserts that societies require self-controlling actors, and that such 'technologies' arise in response to this need. However, this Foucauldian concept is suggested without reference to the part that agency plays in the constitution of social life, and without reference to the contexts of agency (political, occupational, organisational, professional, religious, familial, etc.) and their influence upon the exercise of agency.

Next, we look at the 'cardinal sin' of *reductionism*. It should be emphasised at this point that all the 'sins', or illegitimate forms of theoretical reasoning identified here are not confined to any particular theories or paradigms. Rather, as Sibeon (ibid.: 5) suggests, they 'crop up almost everywhere across the theoretical landscape'. It should be noted that Sibeon refers to *reification, reductionism, essentialism* and *functional teleology* here. Sometimes, as in the work of Hanmer (1993: 232), an author may engage in more than one form of illegitimate reasoning. In this particular case, Hanmer is employing an explanation which relies upon *functional teleology* in the sense of suggesting that 'social forces' (without providing any evidence of conscious planning by social actors) manipulate the new human reproductive biotechnologies, and relying upon *reification* in the sense of suggesting that 'the state' formulates and acts upon decisions pertaining to which women are 'deemed to be unacceptable for motherhood'.

Reductionism

In Sibeon's original framework, and in the modified framework employed here, we regard a *reductionist* theory as one which 'illegitimately attempts to reduce the complexities of social life to a single, unifying principle of explanation or analytical prime mover' (Hindess, 1986a, 1988; Sibeon, 2004: 2). This might be 'globalisation', 'the risk society', 'patriarchy' or 'capitalism'. It must be emphasised here that *reductionism* differs from *essentialism* in that the latter 'sin' is a form of theorising which presupposes a unity of social phenomena. *Reductionism* also differs from *genetic fatalism*, which is a 'new' term included in the framework employed here, *not* employed by Sibeon, and which refers to a tendency to equate genetic predisposition with *inevitability* (in other words, accounts which ignore the evidence, supplied by those such as Ridley (1999, 2003), that genes can be 'switched on' by environmental stimuli). As Sibeon suggests, a key feature of *reductionist* general theories is an 'ontological inflexibility' in which they rest upon 'a priori assumptions' (ibid.), and offer unitary, substantive explication. However, we need to be cautious in our approach to the term 'reductionism'. As the author suggests, to reject 'a priori theoretical commitment to analytical prime movers' is not to suggest that there are 'never situations where a very small number of factors (or perhaps only one) may have causal primacy' (ibid.: 3). However, we should regard this as an empirical question rather than as something (in advance) which is predetermined on the basis of reductionist, unitary explication. Additionally, Sibeon stresses that we should note the difference between 'non-reductive multi-factorial explanation' and what he has termed 'compounded reductionism' (ibid.). The latter form of theorising involves 'attempts to combine or synthesise two or more reductionisms' such as 'capitalism' and 'patriarchy' (ibid.). Thirdly, we are entreated to avoid what Sibeon terms 'deferred reductionism', in which reductionist explication is 'postponed or deferred' (rather than removed from analysis) (ibid.). The author gives the work of Farganis's (1994: 15–16) feminist critique of postmodern theory, in which Farganis 'acknowledges that social class, age, ethnicity, and sexuality are important dimensions' of social life/existence, but argues 'in the last instance' it is always the case that gender is 'the ultimate or primary axis of social life' as a cogent example of 'deferred reductionism'.

Globalisation is a concept which is often drawn upon in contemporary criminology, especially in relation to security studies and the phenomena of globalised crime. However, as Owen (2009b) suggests, this has unfortunately often involved the illicit use of reductionism. In a study which examines the current 'state of play' (circa 2009) and recent trends in the criminal justice system of England and Wales in the context of the impact of the processes which have come to be referred to as 'globalisation' in terms of crime, citizenship, the welfare state and challenges to the nation-state, Owen contends that whilst there is little doubt that the criminal justice system and the individual nation-state of the United Kingdom are being challenged by global and technological processes, we should strive to avoid reductionist theoretical accounts relying upon unitary explanations for complex social phenomena which serve to exaggerate the scale and/or the intensity of globalisation, and which underplay the uneven impact of globalising tendencies. It is the view here that the social world is contingent and not determined by macrostructural social forces.

Recently, Muncie (2007: 186–7) has defined globalisation as 'a widely, but often loosely, used term which usually implies an increasing homogeneity of national economies, politics, and culture'. He then observes that such convergence is driven mainly by 'international flows of de-regulated capital, information and people and dominated by multi-nationals, neo-liberal economies and technologies' (ibid.: 187). Muncie makes the point that global multinational corporations and financial markets 'now seem to provide the economic, political, and cultural parameters in which we now live' (ibid.). For Muncie, 'the sovereignty of individual nation-states and the authority of traditional social institutions seem to be increasingly redundant in the face of these powerful forces' (ibid.). Nelken (1997) has made similar observations. However, as Muncie goes on to suggest, despite a widespread acceptance of these notions, the meaning and implications of globalisation are the subject of great debate. The concept is sometimes used interchangeably with other competing macroconcepts, such as *transnationalisation* (the dissolving of national boundaries); *supranationalisation* (transcending national limits); *internationalisation* (exchanges of capital and labour); *universalisation* (the spread of information and cultural phenomena world-wide); *neoliberalisation* (the removal of regulatory barriers to international exchange/transfer); *westernisation* (standardisation driven by advanced industrial economies); *Anglo-Americanisation* (homogenisation driven by the United States/United Kingdom alliance); or indeed *modernisation* (the diffusion of managerial economies).

As Muncie (ibid.) appears to suggest, dispute emerges over the question of whether globalisation is 'anything new at all, or rather simply a modern version of *colonization'*. Yeates (2001) has observed that the concept is flawed because it encourages reductionist and economistic readings of societal change. Is the concept able to aid an understanding of contemporary transformations in crime and crime control within the criminal justice system of England and Wales? Arguably, as Muncie (ibid.) makes clear, there is an 'emergent and growing fear' that global flows of capital, information, and human beings are providing the ideal conditions and opportunities for organised crime to flourish. Criminal enterprises, for example, the Chinese triads, the Russian mafia, and Jamaican Yardies and others are often assumed to have made widespread profits in the trafficking of drugs, arms, human beings, in international pornography, prostitution and international fraud. This 'vision of criminality out of control' (Muncie, ibid.) has arguably dramatically increased since the terrorist attacks of 11 September 2001, together with the widespread fear that crime now lacks boundaries. However, with regard to serious organised crime, it is important to appreciate the 'subtlety, complexity and depth of field of the organization of crimes' (Levi, 2007: 799). In doing so, we should keep firmly in mind that 'many different forms of organization can coexist in parallel, and that to be an "organized criminal" does not mean that one has to be a member of an "organized crime syndicate"'. As Levi (ibid.) observes, There is no Blofeld figure or SMERSH collective organising crime or terrorism worldwide; rather there are 'layers of different forms of enterprise criminal, some undertaking wholly illegal activities and others mixing the legal and illegal depending on contacts, trust, and assessment of risks from enforcement in particular national markets'.

As Owen (2009b) suggests, there is little doubt that individual nation-states are being challenged by global processes but an analysis at merely nation-state level is limited and limiting as regional governments, federation states, international cities, and multiple forms of community governance suggest alternative visions of statehood and citizenship, and offer some alternative paths of access to decision-making on socio-economic issues. Perhaps, as Muncie (ibid.) appears to suggest, global neoliberal pressures are subject to mediation and can only ever be 'one amongst many influences on policy and then its influence may pull and push in diverse ways *at the same time'*. Thus, 'the global', 'the national' and 'the local' are not exclusive, discrete entities, and the key issue is how they interact and are experienced differently in different spaces and at different times. Loader and Sparks (2007: 91) too appear to acknowledge that globalisation 'is not merely an "out there" phenomenon, a process impacting only on distant occurrences and relations between states'. Its effects are also 'experienced *by* and felt *within* localities that can no longer insulate themselves from events and processes happening elsewhere' (ibid.).

Although it is essential to acknowledge the challenges to the nation-state posed by global processes, it is contended here that we should avoid falling into the trap of underestimating the significance of the nation-state in any analysis of the criminal justice system of England and Wales. Without denying that senses of nationhood or of 'global order' are socially constructed, 'we should not entertain exaggerated claims that the nation-state is no longer a significant entity' (Sibeon, 1996: 149). There is some support for this view from Anderson (1991: 3), who has also observed that 'nation-ness remains the most universally legitimate value in the political life of our time'. Arguably, we should strive to avoid tendencies to exaggerate the scale and/or the intensity of globalisation, attempts to underplay the 'highly uneven impact of globalizing tendencies', and accounts which imply that 'objective and irreversible globalizing forces are at work' (Sibeon, ibid.: 153). In the latter author's view, this orientation leads towards an unfortunate reductionism and essentialism, 'insofar as it erroneously assumes the existence of a social process (globalization) that is relatively unified' (ibid.). Bretherton (1996: 12) has made similar observations, suggesting that 'Globalization is...a set of overlapping processes that are neither inexorable nor irreversible, the impact of which varies in intensity and is highly differentiated in effect. Simply put Globalization is an uncertain process that affects some people more than others.' Not all regions of the world are involved to the same extent in global production, and some such as Sub-Saharan Africa are barely involved at all (Bretherton, ibid.: 7). In an attempt to avoid the 'crude and exaggerated Globalization approach', Sibeon (ibid.: 158) recommends a 'post-national' perspective and methodology which focuses on the subnational, national and transnational levels of governance, rather than 'giving causal primacy to any one of these levels on a priori grounds'. Questions like which of these levels is the most important and whether and in what form there are linkages between them are empirical questions to be determined in each instance, rather than matters for theoretical predetermination in advance of empirical enquiry.

Notions of globalisation may possibly refer to some important economic trends over the past 40 years, but they are limited analytical tools with which to analyse phenomena such as the criminal justice system, globalised crime and criminological issues such as security and terrorism. For example, to conceptualise the complex social processes which have impacted and continue to impact upon the agencies of the criminal justice system, it is necessary to reject 'crude', reductionist, essentialist, unitary explanations, such as those offered by Bell (1973), which effectively reduce the great complexity of social life to a *single* substantive explanatory principle, such as the 'post-industrial society'. It may well be the case that globalisation carries with it transformations which are relevant to the study of crime and criminal behaviour (Muncie, 2007), in the sense of a degree of convergence in terms of criminal justice policies. But as Cavadino and Dignan (2006) have suggested, the adoption of similar economic, social, and criminal justice policies by governments does not necessarily mean uniformity and homogeneity. There may be evidence to suggest that there is a greater cross-jurisdictional awareness among governments of the need to cooperate to deal with crime problems that are not restricted within national boundaries, for example, the UK/USA Drugs Agreement of 1988. In the aftermath of 9/11, it may well be the case that international cooperation is increasing between the member countries of Interpol. It may also be the case that there have been shifts in the nature of citizenship and that views on 'active citizenship' (Walklate, 2007) have taken strong roots in the contemporary political landscape of the United Kingdom and are enshrined in the criminal justice system. However, the determinist grand narratives of 'the logic of post-industrialism' and 'globalisation' are highly contestable. A more adequate and sensible model of social change would arguably recognise that the impact of macro-social processes pertaining to globalisation are most likely to be highly variable and differentially shaped by a variety of economic, cultural and political variables at the subnational, national and transnational levels of social process. We should strive to avoid accounts which make exaggerated claims that the nation-state is no longer a significant entity. As Anderson (1991) has put it, 'nation-ness' remains a universally legitimate value in contemporary political life. Any adequate analysis of, for example, the current 'state of play' in the criminal justice system in the so-called age of 'globalisation' must surely also recognise the role played by agency: patterns of social life and the reproduction of social change are in varied ways influenced by human social actors whose ways of thinking and formulations of interests/purposes are not structurally predetermined nor guided by inexorable 'globalised' motor-forces of change.

In what follows, we apply the meta-concept of *genetic fatalism* to the study of crime and criminal behaviour, in order to demonstrate the explanatory potential of this 'new' term. Sibeon's original anti-reductionist framework does not include this concept, which as Owen (2006a, 2006b, 2007a, 2007b), Owen and Powell (2006) and Powell and Owen (2005) define as the tendency within social science to associate genetic predisposition/determinism with *inevitability*. As we have observed in Chapter 3, Ridley (1999: 53) has shown how there is a *mutuality* and *plasticity* between genes and environment, and how it is a mistake to deem genes as 'immutable' because they 'need to be switched on, and external events – or free-willed behaviour – can switch on genes'.

Genetic Fatalism

As previously stated, this meta-concept refers to the tendency to equate genetic predisposition/determinism with a fatalistic *inevitability*. We have examined evidence from the work of Ridley (1999, 2003) against this 'cardinal sin' of illegitimate theoretical reasoning, which suggests a great degree of *plasticity* and *mutuality* between genes and environment. Here, it is relevant to recall the 'vigorous counter-attack against the idea of genetic influences on behaviour' led by Richard Lewontin and Stephen Jay Gould referred to by Ridley (1999: 306) and mentioned in Chapter 3. As Ridley notes, a favourite slogan of those involved in the counter-attack against the ideas of E.O. Wilson was 'Not in our genes!' Perhaps at the time it was 'a plausible hypothesis to assert that genetic influences on behaviour were slight or non-existent' (Ridley, ibid.). However, the work of Rose et al. (1984) may be 'no longer tenable' in the light of studies in behavioural genetics which offer cogent evidence that 'genes do influence behaviour' (Ridley, ibid.). It will be recalled that in Chapter 3 we have examined the compelling evidence from authors in the fields of evolutionary psychology and behavioural sciences, such as Hamer and Copeland (1999), Enard et al. (2002), Lai et al. (2001), Dunbar (1996), Pinker (1994), Harris (1998), Bogaert and Fisher (1995), Curry (2003) and Cosmides and Tooby (1997) amongst others for a, at least in part, genetic basis for human behaviour.

As Ridley (ibid.: 307) argues, there is no 'escape from' determinism by 'appealing to socialisation', because 'either effects have causes or they do not'. As the author suggests:

> If I am timid because of something that happened to me when I was young, that event is no less deterministic than a gene for timidity. The greater mistake is not to equate determinism with genes, but to mistake determinism for inevitability. Said the three authors of *Not in our genes*, Steven Rose, Leon Kamin and Richard Lewontin, 'To the biological determinists the old credo "you can't change human nature" is the alpha and omega of the human condition'. But this equation – determinism equals fatalism – is so well understood to be a fallacy that it is hard to find the straw men that the three critics indict.

It should be emphasised here that Ridley (1999, 2003) argues for the concept of 'nature via nurture' here, as opposed to simplistic 'nurture' *or* 'nature' arguments. At no point does Ridley argue for *purely* genetic causality, and in fact acknowledges that, even after the discoveries of evolutionary psychology and behavioural genetics pertaining to the influence of genes upon behaviour, 'environment is still massively important, probably in total more important than genes in nearly all behaviours'.

Schaffner (a Professor of Medical Humanities at George Washington University, USA), writing in 2000, argues for a *'genetic-environmental determinism'*, which appears to be very similar to Ridley's 'nature via nurture' model. However, it could be argued that Schaffner (ibid.: 326) engages in *genetic fatalism* when he describes his paper as being 'critical of *genetic* determinism'. Arguably, the author is making the same mistake that Ridley identifies as equating determinism with *inevitability*. In a sense, Schaffner need not posit *'genetic-environmental determinism'* as an alternative to *genetic determinism*, as Ridley has shown how such genetic causality/determinism 'works' in conjunction with environmental influences in any case. Nature depends upon genes and genes need nurture.

Interestingly, Schaffner draws attention to the considerable amount of philosophical and scientific inquiry directed, over the course of the last hundred years, at issues of reduction and reductionism. Here it should be emphasised that Schaffner (ibid.: 302) refers to *reductionism* in the context of the Human Genome Project, and closely associates the concept with the notion of 'genetic determinism' rather than in the sense that we employ the term *reductionism* (as a 'cardinal sin') in the 'new' framework in relation to theories which attempt to reduce the complexities of the social world to a single, unitary explanation. By 'genetic determinism', Schaffner means that *inherited* genes are taken to be the principal drivers of an organism's behaviour (ibid.). The author examines the extent to which there are purely, or even primarily, 'genetic' explanations of behaviour. Reference here is made to Kupferman's (1992) review of Kandel, Schwartz and Jessel's *Principles of Neural Science* (later published in 1995). The book, which Schaffner (ibid.: 304) calls 'a bible of sorts in neuroscience', begins by making the point that behaviour in all organisms is shaped by the interaction of *both* genes and the environment. The relative importance of the two variables varies, but even 'the most stereotyped behaviour' can be 'modified' by environment and the most plastic behaviour (such as language) is influenced by innate factors (Kupferman, ibid.: 987). As Schaffner (ibid.) suggests, Kupferman goes on to focus on aspects of behaviour that could conceivably be inherited, and on the processes of interaction between genes and environment affecting behaviour. Thus, as Schaffner (ibid.) makes clear, the point about the incompleteness of any exclusively genetic basis for the explanation of behaviour is taken as a major premise in the scientific community. The search for *pure, single gene* genetic explanations for behaviour, as Schaffner (ibid.: 323) suggests, may be conceptualised as the 'strongest of "strong" genetic programs', but the concept is revealed to be incoherent. As the author suggests, 'any behaviour requires an organism acting in an environment' (ibid.). The organism may thus be regarded as the 'product' of many genes which are required to 'build' the nervous system 'integrated within the organism that is productive of behaviour'. However, as Schaffner (ibid.) suggests, the Huntington's Disease (HD) paradigm, 'in which we appear to have a single gene that is productive of HD in all current environments', has a powerful hold on both geneticists' thinking and media/popular thought. The reference Schaffner (ibid.) makes is to the work of Rubinsztein et al. (1996) in the *American Journal of Human Genetics*. However, as Schaffner (ibid.: 318) suggests, 'The Huntington's Disease (HD) paradigm of a (prima facie) single dominant 100% penetrant gene' may be more 'the exception rather than paradigmatic' for the search for such single genes for bipolar disorder and schizophrenia 'has *not* been successful'. One is reminded here of the ill-fated attempt by the Centre for Social, Genetic and Development Psychiatry at Maudsley Hospital, London, in 1994 to isolate a 'criminal gene' (Hopkins Burke, 2009).

As Schaffner (ibid.: 317) shows, it is the case that there have been advances in the molecular genetics of 'a variety of *single gene* somatic disorders'. The prototype may be sickle-cell anaemia which dates back to the 1949–57 years, 'when it was characterised by Pauling and Ingraham as a molecular disease due to a single mutation' (ibid.). In 1983, the author shows, the gene for HD was localised to chromosome 4, 'though the specific locus of this gene was difficult to determine, and had to wait until 1993' (ibid.). Genes for polycystic kidney disease, retinoblastoma and cystic fibrosis 'were mapped in 1985, and this latter gene was localised and cloned in 1989' (ibid.). Recently, as Schaffner indicates, 'genes for two types of inherited colon cancer and two forms of breast cancer have been localised and sequenced' (ibid.).

The genetics of mental disorders, however, 'has not fared as well', though some important/controversial advances have been made in 'the behavioural realm' (ibid.). As Schaffner shows, the finding that there was a gene associated with schizophrenia on chromosome 5 'turned out to be fallacious'. Kennedy et al. (1988) support Schaffner's claim. As Kelsoe et al. (1989) has made clear, a report identifying a simple gene for manic-depressive disorder on chromosome 11 had to be withdrawn from the literature. Schaffner (ibid.) makes the point that the only 'single-gene-behavioural trail' example of which he is aware involves 'a point mutation in the structural gene for monoamine oxidose A' that was associated with a 'vaguely defined' impulse/disorder/aggression phenotype reported by Brunner et al. (1993a, 1993b). Although Schaffner acknowledges the methodologically 'sound' science employed by Brunner (ibid.), the latter does not characterise the mutation as an 'aggression gene'. Brunner (ibid.: 160) comments that 'behaviour should and does arise at the highest level of cortical organisation, where individual genes are only distantly reflected in anatomical structure'. As Schaffner (ibid.: 318) suggests, recent discoveries in personality genetics include 'an identification of a gene, DRD4, for "novelty-seeking", but it should be noted that the group reporting a confirmation of that gene noted that this was only a partial – and not a single gene kind of explanation'. The group, according to Benjamin et al. (1996: 83), suggested that 'DRD4 accounts for roughly 10% of the genetic variance, as might be expected if there are ten or so genes for this complex, normally distributed trait'. As we saw in Chapter 3 (the section 'Incorporating the biological variable'), Hamer and Copeland (1999: 180) suggest that D4DR may not be regarded as a 'promiscuity gene', but it 'determines the way certain brain cells respond to dopamine, which in turn influences a person's reaction to novel stimuli'.

Schaffner (ibid.: 326) closes his paper by suggesting that his investigations have not 'offered any arguments to undercut a combined (and historically-based) *genetic environmental* determinism'. This appears to be a similar position to that employed by Ridley (1999, 2003) and favoured in the 'new' modified framework employed here. In Chapter 3 (the section entitled 'Incorporating the Biological Variable'), we have seen how Newton (2003: 29), similarly to the work of Rose et al. (1984), appears to engage in *genetic fatalism* in his treatment of Walter Cannon's (1914) work. Newton (ibid.) also appears to dismiss previous work by Benton (1991: 17) which suggests that Cannon sought to conceptualise life-forms 'holistically'. Newton (ibid.) instead argues that Cannon's arguments are based upon 'reductionistic' accounts which explain human behaviour, 'chiefly in terms of instincts'. As we saw previously, Cannon's (ibid.: 264) suggestion that 'the emotion of fear is associated with the *instinct* for flight, and the emotion of anger or rage with the *instinct* for fighting or attack' is arguably less of a 'reductionistic' and 'desocialised' account than Newton (ibid.) appears to think. It would appear more likely that Cannon is actually arguing for what Ridley (2003: 4) has termed 'Nature via Nurture', a scenario in which genes are reacting to social cues rather than a form of *fatalistic* determinism, in this case the 'cues' being the 'cause' of the emotion of fear.

Wilkie (1994) falls into the trap of equating any explanation which relies upon genes with immutability and *inevitability*, as Owen (2009a) illustrates. The author refers to how, in his view, the rise of environmental movements since the 1960s has a 'bearing on the attractiveness of biological (which almost without exception means genetic) explanations for human conduct' (ibid.: 181). Wilkie (ibid.) refers to how 'sociobiology' appeals to a kind of romanticism which in turn appeals to nostalgia for 'mythical' ideas pertaining to harmony between 'human biological nature' and cultural form in its attempts to explain modern social life. Wilkie (ibid.: 181–2) goes on to argue that 'some proponents of the Human Genome Project might protest that such a borrowing of genetic explanations of human traits and behaviour is illegitimate', but on the other hand, a few of the project's 'foremost proponents' such as Walter Gilbert (a Nobel Prize-winning geneticist) have remarked that 'when we know the complete human genome we will know what it is to be human'. Wilkie (ibid.: 182) contrasts the difference between 'assumptions underlying sociobiology' and the Human Genome Project (we know, of course, that the human genome was sequenced in 2001). The difference lies in the 'alterable' nature of the composition of the human genome as perceived by 'geneticists' and the sociobiological view that 'the iron hand of the genes' is 'immutable' and 'fixed' (ibid.). Wilkie, writing in 1994, appears unaware of the 'reforms from within' sociobiology attempted by the ex-Harvard scholars, John Tooby and Leda Cosmides. The work of the latter authors is (particularly their 1997 work *Evolutionary Psychology: A Primer*) drawn upon in Chapter 3, in which we examine the evidence from evolutionary psychology and related fields for a, at least in part, genetic basis for some human behaviour. As Ridley (2003: 245) points out, Tooby and Cosmides, in their chapter 'The Psychological Foundations of Culture' in Barkow et al. (1992) argued that the expressed behaviour of a human being 'need not be directly related to genes, but the underlying psychological mechanisms could be'. So, as Ridley (ibid.) suggests, Tooby and Cosmides could eschew the idea that we can 'search for "genes for war"', and also 'the contrary dogmatic insistence that war is a pure product of culture written on the blank slate of impressionable minds'. Rather, the authors argue for 'psychological mechanisms in the mind, placed there by natural selection acting in the past upon sets of genes that predispose most people to react to some circumstances in warlike ways' (ibid.). This, Tooby and Cosmides (1992) christened 'evolutionary psychology'. As Ridley (ibid.) makes clear, it is arguably an attempt to 'fuse the best of Chomsky's nativism – the idea that the mind cannot learn unless it has the rudiments of innate knowledge' with the 'best of sociobiology's selectionism: that the way to understand a part of the mind is to understand what natural selection designed it to do'. For Tooby and Cosmides (ibid.), the entire developmental programme thus evolves, and each programme (for an eye, a kidney, etc.) requires the integration of perhaps thousands of genes, but crucially the presence of 'expected environmental cues' (Ridley, ibid.). This, then, may be viewed not as the 'immutable', 'fixed' ideas of 'the iron-hand of the genes'– the *genetic fatalism* fearfully referred to in Wilkie's (1994: 182) account of the implications of the Human Genome Project – but rather 'a subtle mixture of nature and nurture that studiously avoids pitting the two in opposition to each other' (Ridley, ibid.). It should be emphasised here that this approach, termed 'nature via nurture' by Ridley, is the one which was argued for in Chapter 3, and the one which it is suggested here is best suited to be combined with the meta-concepts of Sibeon's (2004) original framework. *Genetic fatalism*, as has been demonstrated here, is a 'cardinal sin' which needs to be studiously avoided in theoretical analysis pertaining to crime and criminal behaviour and related areas involving possible linkage between life and social sciences, and it refers to a widespread tendency within Criminology and other, related social sciences to equate genetic determinism with inevitability. It is a mistake to view the genes involved in human behaviour as immutable. Genes can be switched on, and both external events and free-willed behaviour can switch on genes. As Owen (2012a) suggests, 'This is quite a different concept from the Lombrosian idea that individuals are "born to be criminal".' Together with the *oversocialised gaze* and the *biological variable, genetic fatalism* is a 'new' meta-concept which, as we have seen at the beginning of this chapter, forms part of the 'new' framework. It is to the concept of the *oversocialised gaze* that we must now turn our attention. Again, we examine the explanatory potential of the meta-concept in relation to the study of crime and criminal behaviour.

The oversocialised gaze

The oversocialised gaze refers to accounts which are harshly 'environmentalist' and antipathetic towards genetic or partially genetic explanations of human behaviour. The accounts may even reject genetic variables altogether, as Giddens (1993: 57) appears to do when suggesting that 'human beings have no *instincts* in the sense of complex patterns of unlearned behaviour'. As we have seen in Chapter 3 (the section entitled 'Incorporating the Biological Variable'), there is cogent evidence for *instinctive*, unlearned patterns of human behaviour in the work of authors such as Ridley (1999, 2003), Hamer and Copeland (1999), Pinker (1994) and Dunbar (1996), especially regarding the idea that 'grammar is innate' (Ridley, 1999: 104).

Also examined in Chapter 3 (the section entitled 'Duality versus dualism revisited') was the tendency of writers of the 'embodied school' or 'sociologists of the body' to engage in *oversocialised* approaches. Tim Newton's (2003: 29) criticisms of Sahlins' (1972) and Pollock's (1988) approaches (in the area of 'building bridges' between the social and life sciences) centre around the latter authors' respective portrayals of stress as the product of 'natural' instincts. In Newton's view, 'such arguments present a *crude dualism*' which reduces complex social 'problems' to 'outmoded' conceptions of 'the biological body' (ibid.). He then goes on to argue for a 'putative non-reductionistic "biological sociology"'. As noted in the section of Chapter 3 referred to above, it is hard to envisage a credible 'biological sociology' in the age of the Human Genome Project which denies human instincts altogether. It is not appropriate here to rehearse all the evidence provided in Chapter 3 for the *instinctive* component of human language ability, sexuality and so on. However, it could be argued that Newton (ibid.) appears to be unaware of the fairly recent evidence provided by Ridley (1999) for an account of stress in human beings which posits the *'nature via nurture'* approach to causality as favoured in the 'new' metatheoretical framework applied here. Newton's arguably *oversocialised* concept of stress emphasises the influence of complex social factors whilst rejecting any reference to 'natural', instinctive, genetically based behaviour. Ridley (ibid.) arguably shows how it is indeed possible to include both genetic *and* environment variables in an explanation of stress in human beings.

As Ridley (ibid.: 149) acknowledges, stress in human beings is 'caused' by 'the outside world', and 'short-term stressors cause an immediate increase in epinephrine and norepinephrine', which are the hormones that make the human heart beat faster, the feet go coldand so on. These hormones act to prepare the human body for 'fight or flight' in emergencies. Stressors that last for longer activate a different pathway that results in a much slower but more persistent increase in cortisol. Cortisol is 'used in virtually every system in the body' and is a 'hormone that literally integrates the body and the mind' by altering the configuration of the brain (ibid.). As the author suggests, 'one of cortisol's most surprising effects is that it suppresses the working of the immune system', and those who have shown the symptoms of stress are 'more likely to catch colds and other infections', because one of the effects produced by cortisol is the reduction of white blood cells (ibid.: 149–50). Ridley goes on to show how the relationship between genes and environment is one of *mutuality* rather than an *oversocialised* conception which rejects genes and instinct in favour of a solely social explanation of the kind suggested by Newton (2003) in his treatment of the work of Sahlin (1972) and Pollock (1988). He (ibid.: 150) explains how the hormone cortisol reduces the 'activity, number and lifetime of lymphocytes', or white blood cells, showing how cortisol does this by switching genes on:

> It only switches on genes in cells that have cortisol receptors in them, which have in turn been switched on by some other triggers. The genes that it switches on will then switch on other genes and so on. The secondary effects of cortisol can involve tens, or maybe even hundreds, of genes. But the cortisol was only made in the first place because a series of genes was switched on in the adrenal cortex to make the enzymes necessary for making cortisol.

As Ridley suggests, 'you cannot produce, regulate and respond to cortisol without hundreds of genes', practically all of which 'work' by 'switching each other on and off'. In white blood cells, 'control is almost certainly involved' in 'switching on' the gene known as *TCF*, on chromosome ten, which enables *TCF* to make its own protein, 'whose job is to suppress the expression of another protein called interleukin 2' which in turn is a 'chemical that puts white blood cells on alert to be especially vigilant for germs' (ibid.). Thus, cortisol suppresses the 'immune alertness of white blood' and renders the person 'more susceptible to disease' (ibid.). The author goes on to pose an interesting question: 'Who's in charge? Who ordered all these switches to be set in the right way in the first place, and who decides when to start to let loose the cortisol?' (ibid.). The answer does not lie with the genes solely, even though 'the differentiation of the body into different cell types, each with different genes switched on' is a genetic process 'at root' (ibid.). Genes are not the 'cause' of stress, and as Ridley makes clear, 'the death of a loved one, or an impending exam, do not speak directly to the genes' (ibid.). They are, in essence, 'information' processed by the human brain. Is the human brain 'in charge'? On one hand, as the author suggests, the hypothalamus of the human brain 'sends out the signal that tells the pituitary gland to release a hormone that tells the adrenal gland to make and secrete cortisol' (ibid.: 151). The hypothalamus takes 'orders' from 'the conscious part of the brain which gets its information from the outside world' (ibid.). However, as Ridley suggests, this 'answer' will not suffice 'because the brain is part of the body'. Instead,

> [t]he reason the hypothalamus stimulates the pituitary which stimulates the adrenal cortex is not because the brain decided or learnt that this was a good way to do things. It did not set up the system in such a way that thinking about an impending exam would make you less resistant to catching a cold.
>
> (ibid.)

In answer to what *did* 'set up the system', Ridley suggests that 'natural selection did', and 'somewhere down the cascade of events that is the production, control and reaction to cortisol' stress-prone people 'must have subtly different genes from phlegmatic folk' (ibid.). Who or what is ultimately 'in charge'? Ridley's answer is that 'nobody is in charge', rather an 'intricate, cleverly designed and interconnected system' in which nurture depends upon genes, and genes require nurture in an elegant mutuality. This is, arguably, far removed from the *oversocialised* explanation for stress posited by Newton (2003), or that which the author identifies as '*crude dualism*' in the work of Sahlin (1972) and Pollock (1988). As Owen (2012a: 84) suggests, there are some very important implications here which inform attempts to construct *genetic–social* criminological theory. For example, 'Hostile behaviour can be induced in humans by increasing plasma levels of norepinephrine, whereas agents that block norepinephrine receptor cells can reduce violent behaviour' (ibid.). The enzyme monoamine oxidase is involved in the reduction of norepinephrine, and low levels of the enzyme correlate with violent criminal behaviour as low levels of monoamine oxidase allow norepinephrine levels to increase (Klinteberg, 1996).

In Chapter 3 (the section entitled 'Incorporating the Biological Variable'), we saw how Owen (2006a) has identified the work of Gagnon and Simon (1973) and Michael Foucault (1980b) as examples of accounts which rely upon the *oversocialised gaze*. As Owen (2006a: 190) suggests, 'Foucault's (1980b) argument that sexuality is a socio-cultural creation, that sexuality as we know it is the production of a particular set of historical circumstances and obtained only within the terms of a discourse' is *oversocialised*. Owen refers to the work of Hamer and Copeland (1999: 163), which shows how genes influence our sexual desire, how often we have sex, and 'help make us receptive to the social interactions and signs of mutual attraction that we feel instinctively'. We have examined the genetic evidence pertaining to how the D4 dopamine receptor gene influences male sexual behaviour, from the work of Hamer and Copeland (ibid.), in the section of Chapter 3 previously referred to. There is thus no need to rehearse the argument here. In the case of Foucault's (1980b) work, his belief that sexuality is 'controlled' through definition and regulation (rather than by the 'impression' suggested by Adrienne Rich's (1981) feminist analysis of heterosexuality), in particular via the creation of sexual categories such as 'heterosexuality' and so on, is open to question. In the case of female sexuality, Foucault (ibid.) argues that it is 'controlled' not by the denial or 'silencing' favoured in Walby's (1990) feminist account but by *constant referral*. Foucault suggests that the 'history of sexuality' is in effect a history of shifting forms of control/regulation. In the case of the last hundred years or so, the shift has been away from the church's moral regulation towards an increased regulation through education, medicine, psychology, law, social work, social policy and so on. 'Sex', for Foucault (ibid.), is not a biological 'entity' but rather an *idea*, a concept which is always specific to certain cultures and cultural and historical periods. Sexuality for Foucault is produced through *discourses on sexuality*, which shape human sexual values and beliefs. This, as Owen (2006a, 2006b, 2007a) suggests, appears to be an *oversocialised* perspective which denies the notion of a biological sexual drive altogether. As Ridley (1999: 149) shows, 'a gene on chromosome 10 called CYP17' directly affects sexuality in human beings. The gene manufactures an enzyme which enables the human body to convert cholesterol into cortisol, testosterone and oestradiol, and without the enzyme, the pathway is blocked and the only hormones that can be made from cholesterol are progesterone and corticosterone. People lacking a 'working copy' of this gene 'cannot make other sex hormones so they fail to go through puberty; if genetically male, they look like girls' (ibid.).

Owen (2006a) has identified Gagnon and Simon's (1973) symbolic interactionism as *oversocialised*. The authors suggest a very radical form of social constructivism in which there is (as is the case with Foucault (1980b)) no sexual drive in the 'natural' sense that it is said all human beings possess to varying extents as part of biological make-up. The authors famously argue for a model of human sexual drive purely as social construction. Not only do we *learn* what 'sex' means, and who/what is sexually arousing to us, but we also learn to want sex. Though the authors acknowledge that the body has a repertoire of biological 'gratifications' (that is, the capacity for orgasm), it does not automatically follow that we would want to engage with them. Certain 'gratifications' will be selected as 'sexual' via the learning of 'sexual scripts'. For example, a particular experience would not be repeatedly sought after unless there was the presence of a 'meaningful' script. From Gagnon and Simon's *oversocialised* perspective, which denies a biological component to *sexual drive* (different from the 'gratifications'), socialisation is not about learning to control innate sexual desires in order that they are expressed in socially acceptable ways, but rather the learning of 'sexual scripts' of some complexity which serve to specify circumstances which will elicit sexual desire and 'make' the person wish to engage in certain acts with his/her body. Thus, for the authors, sexual drive is, as it is for Foucault (1980b), a 'learnt social goal'.

The *oversocialised gaze* is an important meta-concept of the framework, and most definitely an illegitimate form of theoretical reasoning, or 'cardinal sin', which we should avoid when engaging in the study of crime and criminal behaviour and or related areas pertaining to bio-social issues, especially attempts to 'bridge' the social and biological sciences. However, it is equally important to avoid accounts which are *under-socialised*. As the work of Powell and Owen (2005) has suggested, we require a 'balanced' approach to analysis which acknowledges the role of 'the social' alongside the biological in 'biomedical' matters. This is in line with the approach favoured here, which acknowledges a *mutuality* between genes and environment in terms of causality. Following Mouzelis (1991), Powell and Owen (ibid.: 28) suggest that 'sociological' approaches, as opposed to rigid, reductionist 'biomedical' approaches, are required especially in analysis of the relationship between *macro* and *micro* tiers and links, and between temporal spatialities or contingencies between *agency* and *structure*. To recap, the *oversocialised gaze* refers to harshly 'environmentalist' accounts which are characterised by a strong antipathy towards genetic, or partially genetic, explanations; for example, the theories of the Chicagoans, Shaw and McKay (1942), which imply that individual action can be explained solely by the larger environment in which the individual resides, and Sutherland's (1947) view that criminal behaviour is entirely learned. Having examined the explanatory potential of the *oversocialised gaze* as a meta-concept of the 'new' framework, we must next look at the concept of the *biological variable*.

The biological variable

At the beginning of this chapter (in the section entitled 'The Metatheoretical Framework') which essentially involves a codification of the framework/meta-concepts, the *biological variable* is referred to as a 'new' term. The term, or meta-concept, is 'new' in the sense that it is *not* employed by Sibeon. We should regard the meta-concept as referring to the evidence for a, at least in part, *biological* basis for some human behaviour. Its use arises out of a respectful critique of Sibeon's original anti-reductionism on the grounds that it neglects biological or part biological causality in explication of social 'reality' (Owen, 2006a, 2006b, 2007a, 2007b; Owen and Powell, 2006; and Powell and Owen, 2005). As was made clear in the previous sections on other 'new' terms ('*the oversocialised gaze*' and '*genetic fatalism*'), the evidence *for* the '*biological variable*' comes from the literature of evolutionary psychology and behavioural genetics in the work of Ridley (1999, 2003), Dunbar (1996), Hamer and Copeland (1999) and others. The evidence for genetic/partial genetic causality in relation to selected behaviours such as *sexuality* (Hamer and Copeland, ibid.; Owen, 2006a; Bogaert and Fisher, 1995), *language* (Hamer and Copeland, ibid.; Dunbar, ibid.; Enard et al., 2002; Pinker, 1994) and *reactions to stress* (Ridley, 2003; Pollock, 1988) have been discussed in some depth in Chapter 3 and variously in the sections on 'the *oversocialised gaze*' and *'genetic fatalism'*.

It is the contention here that in theoretical analysis, particularly that pertaining to the study of crime and criminal behaviour and related areas in which links are attempted between 'the social' and biology (Benton, 1991, 1994; Williams, 1998, 2003; Bury, 1997; Newton, 2003), it is *essential* to recognise that 'genes do influence behaviour' (Ridley, 1999: 306) and that the harshly 'environmentalist' accounts of Foucault (1980b) and Gagnon and Simon (1973) in which sexuality, for example, is completely *learned* behaviour are no longer plausible hypotheses in the age of the Human Genome Project. Wilkie's (1994: 171–2) treatment of the 'moral consequences of molecular biology' contains the expression of unease at the possibility that the Project will 'point up differences between individual humans at a genetic level'. The author suggests that if such genetic knowledge is 'not handled properly and seen in its proper, biological context', it may lead to the generation of information which enables 'new grounds of discrimination'. From the point of view of 'the Project's very existence', we may take an increasingly ' "atomistic" view of human beings and indeed of life itself' (ibid.). Wilkie's (ibid.) greatest fear is that the advances in human biotechnology may lead to a tendency to 'define ourselves in genetic terms and neglect the rest'. The author draws attention to the possibility of the risk that 'we may all become reductionists reducing our lives to their supposedly fundamental components', missing the holistic 'complexity and richness of life in its entirety'. Arguably, Wilkie's fears are misplaced. There is no reason to suppose that such *'geneticisation'* (also referred to by Opitz, 2000) should ever come about. As we have seen in this chapter's section on 'Genetic Fatalism', Schaffner makes it clear that the incompleteness of any *exclusively* genetic basis for human behaviour is taken as a general premise within the scientific community. The *'biological variable'* should be recognised as playing a part in causality, but equally there is a role for the environment (part of the 'complexity' and 'richness' referred to by Wilkie) which is 'massively important' (Ridley, ibid.), in the sense that genes are 'switched on' by social cues. Acknowledging a genetic component to causality does not have to entail the *genetic fatalism* engaged in by Wilkie. The three authors of *Not in our genes* (referred to in Chapter 3, in the section entitled 'Building Bridges', pertaining to attempts to create theoretical links between 'the social' and biology), Steven Rose, Leon Kamin and Richard Lewontin made the point that 'biological determinists' believe the credo 'you can't change human nature' to be the beginning and end of the human condition. However, as Ridley (1999: 307) makes clear, 'this equation-determinism equals fatalism – is so well understood to be a fallacy that it is hard to find the straw men that the three critics indict'.

In essence, the evidence (in terms of selected examples) for biological variables playing a part in human behaviour has been explored in Chapter 3, and the source of the insights (largely *evolutionary psychological*) has been examined with reference to the staunch defence of such ideas by Curry (2003), Dawkins (1982, 1985), Daly and Williams (1998), Tooby and De Vore (1987), Cosmides and Tooby (1997), Ridley (1999, 2003) and so on, and criticisms by those such as Rose and Rose (2000), David (2002), Rose (2000) and Rose et al. (1984). It will be recalled that the decision to incorporate the idea of a *biological* basis (which does not imply *fatalism* and acknowledges the *plasticity* and *mutuality* embracing genes and environment/social cues for human behaviour) was made at the end of Chapter 3, after reviewing evidence pertaining to genetic causality/partial genetic causality in the cases of selected examples of human behaviour.

Owen (2012b) has recently applied this notion of the biological variable to the study of masculinities and crime, examining competing theories of 'masculinities' in relation to violence and crime. There is currently considerable interest in the field of 'masculinities and crime', as is evidenced by the work of Winlow (2001), Winlow and Hall (2009), and Tomsen (2008). Owen (ibid.) draws upon the work of authors such as Connell (2000), Messerschmidt (1993), Collier (1998), MacInnes (1998), Powell (2006) and Jefferson (2007), who work within the sociological and criminological traditions, considering the evidence for the social construction of masculinities and male violence. It is the contention here that social constructionist theories, such as those associated with Connell (2000), are interesting and useful in relation to the theoretical analysis of masculinities as gender and gendered identities. However, they contain theoretical deficits and shortcomings in relation to the analysis of male sexualities. Here it is suggested that a post-postmodern analysis of masculinities might incorporate some of the insights from the *genetic–social* metatheoretical framework, and in particular the notion of the biological variable. We have already examined the cogent evidence from those such as Hamer and Copeland (1999) which counters the Foucauldian idea that human sexuality, and in this case male sexuality, is merely a socio-cultural creation and the production of a particular set of historical circumstances and obtains only within the terms of a discourse developed since the seventeenth century. In addition, it is recommended that an analysis of masculinities and crime might incorporate Layder's (1997) notion of *Psychobiography*, the unique, asocial aspects of the person, such as disposition, modified within the *genetic–social* framework to embrace the mutuality of relations between genes and environment. We will examine the notion of Psychobiography and its applications to crime and criminal behaviour later in this chapter.

The idea that masculinities are socially constructed goes back to early psychoanalysis, and in social science research first took the shape of a social-psychological concept, the 'male sex role'. Such an approach emphasised the learning of norms for conduct and has been popular in social areas of concern such as educational studies (Connell, 2000). However, sex role theory is inadequate for understanding the power and economic dimensions in gender. Further, 'it is telling that discussions of "the male sex role" have mostly ignored gay men and have little to say about race and ethnicity. Sex role theory has a fundamental difficulty in grasping issues of power' (Connell, 2000: 27). Connell points to an explosion in masculinity studies which focus on marital sexuality, 'homophobic' murders, body building culture, insurance, public and private violence, professional sports, criminal justice and the literary genre. Connell (ibid.) calls such an array of research an 'ethnographic moment' in which the local and specific are emphasised. As Jefferson (2007: 246) has recently suggested, within Criminology, scholars such as Messerschmidt (1993) have utilised Connell's (ibid.) concept of a 'tripartite structure of gender relations and hegemonic and subordinated masculinities as well as the importance of practice', and applied them to contemporary theorising about crime and criminal behaviour. Messerschmidt conceptualises all structures (class, race, etc.) as being 'implicated simultaneously in any given practice, and practices are situationally constrained by the need to "account" for our actions to normative conceptions (of appropriate gender/race/class conduct)' (Jefferson, 2007). Within such a theoretical framework, crime may be conceptualised as a 'resource' for specific men in specific social settings for the 'accomplishment' of masculinity. The significance and importance of crime as a 'resource' for certain men depends upon other 'resources' at their disposal, and these in turn are a 'product of their position in class, gender and race relations and the sorts of situations they find themselves in' (Jefferson, ibid.: 246). As Jefferson makes clear, critics of Messerschmidt's attempts to offer explanations for 'the doing' of all kinds of crime, from varieties of work-based crime to diverse forms of street crime as different ways of 'doing' masculinity, have begun to question whether the author's key ideas are in effect unitary, reductionist explanations.

Some scholars such as MacInnes (1998) view masculinity as an 'ideology' which has developed to enable people to understand continuities of sexual and gender inequalities in what Jefferson terms 'an age of formal equality'. Collier (1998) has sought to develop a relational account of masculinities which acknowledges the 'sexed' body (in common with writers of the 'embodied' school) without 'reverting to a biological essentialism' (Jefferson, 2007: 247). Arguably, as Owen (2012b) demonstrates, it is quite possible to 'bring in biology' in theoretical analysis without resorting to essentialism. Essentialism, employed as a 'cardinal sin' in the *genetic–social* metatheoretical framework, refers to an illicit form of theoretical reasoning that presupposes in aprioristic fashion a kind of unity or homogeneity of phenomena, such as 'white men' or 'middle-class women'. It is possible to utilise the meta-concept of the biological variable in a non-essentialist fashion. The evidence for the biological variable comes directly from the literature of behavioural genetics and evolutionary psychology such as the work of Ridley (1999, 2003), Dunbar (1996), Hamer and Copeland (1999) and others. The evidence for genetic/partially genetic causality in relation to selected behaviours such as sexuality (Hamer and Copeland, 1999; Bogaert and Fisher, 1995), language (Dunbar, ibid.; Enard et al., 2002; Pinker, 1994) and reactions to stress (Pollock, 1988; Ridley, 2003) are discussed in some depth in Owen's (2009a) *Social Theory and Human Biotechnology,* and the meta-concept of the biological variable is incorporated into the *genetic–social* sensitising device/metatheoretical framework. It is the contention here that the meta-concept of the biological variable is of great value and should be incorporated into the theoretical analysis of masculinities and crime. Arguably, accounts of

'Hegemonic masculinity' as favoured by Connell (1995) rest upon an illicit *functional teleology* in the sense of failing to provide evidence of intentional planning by actors 'someplace, sometime', as is the case with Hall et al.'s (1978) *Policing the Crisis*, which similarly draws upon Gramscian concepts of hegemony. Such accounts appear far less convincing than the scientific evidence for the role of cortisol in aggressive criminal behaviour (Martin, 1997; Owen, 2012a). An acknowledgement of the biological variable, for a, at least in part, biological basis for some human behaviours might inform the study of masculinities and crime associated with the social constructionism of Collier (1998), Connell (1995, 2000), Jefferson (2007), MacInnes (1998) and Messerschmidt (1993). To recap, all these scholars provide interesting evidence for the social construction of masculinities. However, they neglect to acknowledge biological variables in their analyses, and their work contains serious theoretical deficits and shortcomings in relation to the role of male sexuality in crime and criminal behaviour, which lies beyond social construction in the realm of genes.

As Owen (2012a) suggests, this acknowledgement of the role of the biological variable in the metatheoretical analysis of crime and criminal behaviour is absolutely essential if we are to point a 'way forward' for contemporary, post-postmodern criminological theory beyond the four main theoretical obstacles that confront the discipline and related disciplines at the present time. As we saw in Chapter 1, these are the nihilistic relativism of the cultural turn, the oversocialised 'bio-phobic' accounts which still dominate mainstream Criminology, the genetic fatalism inherent in theorising which wrongly equates genetic determinism with inevitability, and the sociological weaknesses of many so-called biosocial explanations which appear to neglect or make insufficient use of meta-concepts such as agency–structure, micro–macro and time–space in their accounts of the person. To reiterate, the term *genetic–social* has been employed to refer to the ontologically flexible framework drawn upon and applied here, in order to distance it from hardline Sociobiology. It is the intention here, it will be recalled, to advance upon Shilling's (1993) starting point for a biological sociology, and to supercede the biologically top-heavy, largely American attempts at biosocial analysis (Herrnstein and Murray, 1994; Mednick and Volavka, 1980; Walsh and Beaver, 2009; Walsh and Ellis, 2003; Wilson and Herrnstein, 1985), which appear to lack a sufficiently sophisticated appreciation of sociological theory that would make them truly 'biosocial'. The point has been made earlier that the approach favoured here places the *genetic–social* framework at odds with the left idealist constructionism of Critical Criminology and Abolitionism, which, Jock Young (2013: 252) claims, 'problematizes the way in which crime is defined and constructed yet tends to ignore crime itself'. Whilst Young is arguably correct in his assessment of left idealism, it must also be emphasised that *genetic–social* criminological theorising is also placed at odds with the left realist approach that Young favours. For Young (ibid.), 'the left realist critique argues that explaining crime necessitates a double aetiology of action and reaction and a corresponding denial of any essence of crime'. However, left realism neglects biological variables in analysis just as the left idealism of Critical Criminology and Abolitionism does. Realism is arguably not the 'integrated theory' that Young claims it to be because of this neglect of asocial aspects of the person, such as disposition, and the mutuality of the relationship between genes and environment in criminological theorising and analysis.

In what follows, we apply the meta-concept of a modified notion of Foucauldian power to the study of crime and criminal behaviour in order to demonstrate its explanatory usefulness.

Power

At the beginning of the chapter (in the section entitled 'The Metatheoretical Framework') it is emphasised that the 'new' framework utilises a modified notion of Foucauldian *power* favoured by Sibeon (2004). The author, in a similar position to Derek Layder (2004), acknowledges the *multiple* nature of power. Sibeon's (ibid.: 145) view is that 'power exists in more than one form' – these being *objective, structural* (including *systemic*) forms of power and *agentic* power. The latter meta-concept, *agentic power* refers to what Sibeon (ibid.) considers to be the ability of agents to 'shape events in a preferred direction', and this capacity can be both relational and variable. This modified notion of Foucauldian power has also been employed by Owen (2006a) and by Powell and Owen (2005) in the field of social gerontology/sociological theory. The latter authors make the point that we should reject 'the reductionist, methodological collectivist and macro to micro theories of bio-medicine' because they effectively 'view power as an outcome of their causal powers as top down: in any power the gains of one side (bio-medical mode) dominate the other (older people)' (Powell and Owen, ibid.: 27).

It is the contention here that the meta-concept of *power* (modified, Foucauldian power – modified in the sense that we must recognise the dialectical relationship between *agentic/systemic* forms; the *relational contingent* and *emergent* dimensions; and that power can be stored) has good explanatory potential in relation to notions of 'the state' in terms of criminological analysis, particularly in terms of countering the reified concept of the State favoured by many left idealist 'victimised actor' theorists of crime, who appear to adhere to the Marxist idea of the state as crimogenic, such as Coleman and Sim (2013). It is not only leftist idealism that contains illegitimate conceptions of the state. As we have seen in the section on 'Reification' earlier in this chapter, when Garland (2001: 120) suggests that the punishment of criminals is 'the business of the state', which he regards as symbolic of 'state power', and that 'the state' is responsible for the care of offenders, he is engaging in illicit reification. Arguably, we may view the state in terms of power, as an example of Sibeon's (1996: 52) 'emergent power'. We could analyse the state drawing upon Callon and Latour's (1981) theories (though rejecting their idea that power cannot be stored in positions/roles, etc.), and Foucauldian perspectives, and conclude that power is not a fixed capacity but rather is at least partly *emergent*, in the sense of being an *outcome* of social interaction. Drawing upon the work of John Law (1986a: 5), we could consider power as being, in part, an 'effect' rather than a cause of strategic success achieved by social actors in specific situations. Therefore, in Foucauldian terms, we need to examine the 'genealogy' of existing relations, how they have 'emerged' and the discourses they may reflect. However, we must repudiate the anti-foundational *relativism* inherent in Foucauldian analysis. Is the state to be regarded as an actor in the fashion of left idealism? We need to consider Sibeon's (1996: 57) view concerning the two main (related) points pertaining to social actors when answering that question. For the author, 'they have an ontological status', in the sense that they are able to formulate and act upon decisions using agency, and secondly 'this may be an intermittent status' that does not apply in every circumstance (ibid.). In the case of the state, it is an entity that is *not* an actor. To suggest otherwise would be to engage in *reification*. In some cases, 'the government of a group of "nation states"' may decide that it is desirable to create a project such as the Human Genome Project that is 'empowered' (Sibeon, ibid.: 57) to take decisions/act on the collective behalf of its member states. It is essential, if we are to adhere firmly to the 'rules' of the new framework, to keep firmly in mind Hindess's (1986a: 115) useful formulation of social action, 'a locus of decision and action where the action is in some sense a consequence of the actor's decisions'.

It might prove fruitful to draw upon Bennington and Harvey's (1994) concept of transnational and inter-regional policy networks composed of governmental/non-governmental actors and 'apply' it to the state as a whole. Arguably, we need to acknowledge the multiplicity of forms that *power* takes in relation to the state. This includes *systemic* power 'stored', '*contra* Callon and Latour (1981), in discourses and the social positions/roles of actors, and *agentic* power which refers to a capacity of actors'. We might 'apply' the idea of Sibeon's (2004: 136) *systemic* power, in the sense that doctors and other professionals (aside from being individual social actors capable of 'possessing' power) occupy *roles* and *positions* within which, *contra* Foucault (1991) and Callon and Latour (1981), 'certain elements of power can be stored in positions/roles, social institutions and social systems'. As Sibeon (ibid.) has suggested, Foucault and Callon and Latour 'tend to push their relational and processual conceptions of power to the point of denying that power can be "stored" in roles and in social systems and networks of social relations'. Here, we have adopted Sibeon's (ibid.) synthesis of a combination of Foucauldian and other *relational* concepts of power, with a *systemic* understanding of power. As Sibeon (ibid.) suggests, the synthetic conception of power leads to the idea that 'agentic power nearly always has a relational, contingent and emergent dimension', which is arguably why the position of what Sibeon calls 'top dogs' in any institutional sphere is sometimes 'precariously sustained'.

We therefore should regard as legitimate the idea that an actor possesses power (which invokes the idea of 'storage') as long as we do as Sibeon (ibid.) suggests, and recognise 'how that power is constituted and reproduced relationally'. The author makes a reference to the work of John Law (1991b) here in relation to the latter point.

As Sibeon (ibid.) has made clear, there is good evidence to regard power as 'partly preconstituted and stored' (in roles, etc.), and 'partly relational, emergent and contingent, with the extent to which power is "systemic" or "relational" being an empirical variable' that may vary situationally. This leads to the idea that there are multiple forms of power like agentic, systemic and relational ones as previously mentioned. In the case of power's 'multidimensionality' (Sibeon, ibid.: 137), we must consider the possibility that there is a 'multi-level aspect' to this phenomenon. In other words, there could be 'hybrid', 'more variegated' forms of power than systemic-relational distinctions might imply. As Sibeon (ibid.) suggests:

> Agentic power in some circumstances has a largely systemic source deriving, say, from position/role. In other circumstances agentic power may be of a relatively contingent, emergent kind that emerges during and as an effect of social interaction at the micro or mezo (for example, inter-organisational) levels of social process. Such power may interact with systemic or positional/role power of the type that is 'stored' in discourses, social institutions, and social systems/social networks. There tends, in other words, to be a two-way (dialectical) relation between systemic and relational forms of power, with each to some extent conditioning the other.

Discourse may then, in a sense, be seen to 'embody' power, for example, the discourse around/about the aetiology of crime. The discourse 'stores' power which relates to multiple, countless decisions taken by professionals and administrative actors that, in turn 'shape' the structure within which the professionals such as judges, magistrates, police officers and criminologists operate and influence agency/decision-making. However, we must apply great caution here. Discourses themselves are *not* actors. *Contra* Foucault, discourses are (in the sense that they are patterned ways of thinking and behaving) 'a form of material that must be mobilised by *actors*' before the discourse can be regarded as having 'any social consequences or effects' (Sibeon, ibid.: 72). As Owen (2012a: 92) suggests, Garland's (2001) 'crime control' thesis has an 'over-reliance upon the Foucauldian concept of discourse, which appears to rest upon a reified notion of agency'. Like Foucault, David Garland appears to presume that discourses are themselves social actors in the sense of a 'crime consciousness' diffused through the media. To reiterate, discourses are a form of material that needs to be manipulated by social actors. In *genetic–social* theorising, we may conceptualise discourse as a potential influence upon social actors, but to regard discourse as an agent is a 'cardinal sin' of reification.

Of great relevance here is Derek Layder's (1997, 2004) multi-dimensional conception of power. Sibeon (2004: 187) notes how Layder's convincing attempt to 'construct a multi-level research strategy' has 'a certain amount of affinity' with his own. Layder's stratified social ontology is, as Sibeon (ibid.) notes, 'a modification and elaboration' of the *distinction* between 'micro-, mezo- and macro- domains'. Layder (1997), as we have previously noted in Chapter 2 (in the section devoted to the author 'Derek Layder'), identifies four *social domains* or 'layerings' as he has sometimes called them. These four *domains* have been defined and discussed in the aforementioned section of Chapter 2. However, to briefly recap, they may be defined as follows. *Psychobiography* refers to largely unique, asocial components of self and behaviour. *Situated activity* refers to face-to-face interaction and intersubjectivity in situations of co-presence. *Social settings* are the locations in which situated activity occurs. *Contextual resources* relate to divisions of class, gender and ethnicity, consisting of macro-distributions and ownership of resources. Each domain has relative autonomy, but they may overlap and influence each other. As Sibeon (ibid.) makes clear, any connections developing between them must be viewed as 'loose', lacking direct links between the micro, mezo and macro tiers of his own research strategy, or in Layder's terms, the four domains (psychobiography, situated activity, social settings and contextual resources). Power for Layder (1997: 13, 250–1) concerns many forms and is 'objective' or external to individual subjectivity, embodied in discourse (which Foucault, (1980b) acknowledges, and Sibeon (2004) appears to agree with, in the sense that power can be 'stored'), but can also be 'subjective'/intersubjective. Sibeon (ibid.: 184) interprets Layder's (ibid.) conception of power as having 'relatively independent, distinct and domain-specific modes of existence'. Power 'wends its way across the micro-, mezo- and macro-domains that constitute social life' (ibid.). Power too, like the four domains Layder identifies, may be interpenetrative and overlapping, with more or less loose or indirect links between the different forms that power assumes either in terms of the asocial aspects of the person, the face-to face interaction, the locations in which interaction/intersubjectivity occurs, or the divisions of class, gender, ethnicity and so on.

If we examine the state in terms of Layder's *social domains* theory, we might view power as 'stored' in the domains of the individual *psychobiography* of key actors; in the face-to-face interaction and intersubjectivity involving the co-presence of actors in their various roles and positions; in the *social settings* in which *situated activity* occurs; and in the *contextual resources* pertaining to divisions of class, gender and race among actors. Power would also be seen as 'embedded' within the discourses manipulated by actors within the layers of the state. As we have seen with reference to a criticism of Garland's (2001) over-reliance upon relational and processual Foucauldian concepts of power, despite the theoretical deficits identified, aspects of Foucault's conception of power may be successfully incorporated into *genetic–social* analysis of crime and criminal behaviour. This must be done in a critical and selective fashion, informed by the critique of agency–structure, micro–macro and other terms offered in the 'new' framework. If we are to employ Foucauldian insights pertaining to power in criminological theorising, we must recognise the dialectical relationship between agentic and systemic forms of power; the relational, contingent and emergent dimensions of power; and the concept that, *contra* Foucault, aspects of power can be stored in positions/roles such as those of police officers and magistrates, and in social systems/networks such as the criminal justice system.

Having demonstrated the explanatory potential of the 'new' framework's modified notion of Foucauldian power, we next move on to the usefulness of the concept of *dualism of structure* in studies of crime and criminal behaviour and related areas involving links between the social and biological sciences.

Dualism

In Chapter 2 (the section entitled 'Agency and Social actors') we discussed the Giddensian rejection of dualism in favour of a duality of structure. Later in the same chapter (in the section entitled 'Archer, Layder and Mouzelis: Post-postmodern Theorists') we discussed the approaches to the question of 'dualism or duality' in relation to the theoretical and research strategies of the aforementioned major sociological thinkers. Chapter 3 saw further and more involved discussion of the question, this time pertaining specifically to biosocial issues such as the attempts to 'build bridges' between the life sciences and 'the social' by authors of the 'embodied school' or 'sociology of the body', such as Williams (2003), Benton (2003), Newton (2003) and others in 'Duality Versus Dualism Revisited'. We saw that most 'material-corporeal' sociologists appear to be very much united in a desire to 'challenge dualism on all fronts' (Williams, ibid.: 556). Benton (2003: 292), it will be recalled, argues for an approach which would 'dissolve dualistic oppositions' (nature/society).

To recap, as we have seen at the beginning of this chapter, *dualism* is favoured because it is felt that in social analysis, *agency–structure* and *micro–macro* should be employed as dualisms that refer to distinct, relatively autonomous phenomena. We seek here to avoid what the 'embodied' writer Shilling (1993) argues for in his analysis of the human body as 'simultaneously' biological and social. Here, Shilling (ibid.) has arguably collapsed both terms into an amalgamated whole within which elements cannot be separated. This is a similar position to Giddensian *structuration*, and the Foucauldian (1972, 1980a, 1980b) tendency to compact *agency* and *structure* together. As Owen (2006a) has suggested, if we collapse distinctions between either *agency–structure* or *biology – the social*, it results in what Archer (1995) calls 'central conflation'. This emphasis upon dualism, it should be emphasised, reflects the influence of the work of Roger Sibeon and Derek Layder upon the construction of the framework. Both authors favour *dualism* rather than the elisionist tendencies of Giddens. Sibeon (2004: 98) makes the point that a 'failure to acknowledge temporality' will result in an 'inability to conceptually separate agency and social structure in a way that might allow us to study links between the two and allow us to assess the reflective influence of each in any given social context'. This is arguably correct, and applies equally to *biology* and '*the social*'.

It will be recalled that we favour here a conception of the relationship between genes and environment, in relation to causality, that Ridley (1999, 2003) calls 'Nature via Nurture'. In other words, we acknowledge that there is a 'feedback loop' that embraces genes and environment, a mutuality which means that genes predetermine the broad structure of the human brain and absorb formative experiences, and react to environmental 'social cues'. This does not mean, however, that we cannot analyse genes and environment, biology and the social separately. Maintaining *dualism* as a meta-concept enables us to do so. We need to acknowledge that genes and environment *both* play a role in human behaviours and *dualism*, if correctly employed in criminological theorising is a useful concept which enables the separation of elements like agency, structure, biology, 'the social' and so on, enabling us to examine the links between them and to assess the relative influence of each element in any social context. In what follows, we apply the meta-concept of Psychobiography to the study of crime and criminal behaviour.

Psychobiography

It is the contention here that there are close theoretical links between Owen's (2009a, 2012a) concept of the biological variable and Derek Layder's (1997) cogent, useful concept of Psychobiography; the largely unique, asocial components of an actor's disposition, behaviour and self-identity. These aspects are regarded by Layder as relatively independent of face-to-face interaction and the macro-social. For Layder, human beings are composed of unique elements of cognition, emotion and behaviour that are, in some sense, separable from the social world, whilst at the same time related in various ways to social conditions and social experiences. Arguably, as is suggested by Owen (2012b), it would prove useful to incorporate notions of unique, asocial Psychobiography into the analysis of masculinities, and in particular male sexuality, in relation to violence and crime. As Layder suggests, asocial elements are separable from but yet linked to the social world. This form of theoretical reasoning appears to have much in common with the *genetic–social* approach favoured here in the sense that it acknowledges the mutuality of the social and the asocial. Derek Layder, like Roger Sibeon, Nicos Mouzelis, Margaret Archer and the author, appears to be advocating to some extent a post-postmodern renewal of sociological theory and method, favouring a flexible ontology which avoids both the absolutist knowledge-claims of meta-narratives and the reductionism and essentialism of modernist paradigms. Layder appears to favour a cogent, 'modest' approach to social explanation, which retains a distinct epistemological commitment to realism, recognising that society is multiform, relatively indeterminate and difficult to predict. This metatheoretical approach is similar to Sibeon's original, anti-reductionist framework, in that it avoids unitary, reductionist explanations and opposes the idea of duality of structure, indeed any attempts to collapse distinctions between agency–structure, micro–macro and so on.

Recently, Owen (2007b, 2012a) has drawn upon Layder's notion of psychobiography in relation to David Garland's 'Culture of Crime Control' thesis, and in particular, 'Garland's rather under-theorised concept of "the individual"' (Owen, 2007b: 9). Here, it is suggested that Garland's implied call for a recognition of criminals as 'individuals' would be strengthened by a recognition of the individual-subjective referred to as Psychobiography by Layder, and the intersubjective, which is labelled Situated Activity. Arguably, these insights should be incorporated into the analysis of masculinities in relation to violence and crime in order to avoid reductionist, oversocialised concepts of male identity and male sexuality. Encouragingly, in the foreword to Owen's *Social Theory and Human Biotechnology* (2009a), Layder himself writes that:

> Owen points out a connection with my own notion of 'psychobiography'. By this concept I originally intended to emphasise the relatively autonomous and unique psychological aspects of individuals' predispositions and behaviour that interact with social influences to produce emergent effects. However, I fully concur with Owen's 'extension' of the implications of the notion of psychobiography to embrace the mutuality and plasticity of the relations between genetic and environmental influences.
>
> (2009a: xiv)

It is possible that this modified notion of Psychobiography might contribute towards the psychological profiling of serial criminal offenders. McLaughlin (2013: 404) refers to the phenomena of serial killing as 'Stranger-perpetuated murders- usually by men- which often appear motiveless and are characterized by gratuitous violence'. Mercifully, such murders are relatively rare but they 'stand at the apogee of a popular fascination with crime, fuelling both a robust mythology and demands for retribution to be at the heart of criminal justice' (ibid.). Criminologists appear to have reached a broad consensus that serial killers differ from 'normal' single incident murderers and other forms of multiple killers such as *'mass murderers'* and *'spree murderers'*. As McLaughlin (ibid.) makes clear, the serial killer tends to exhibit certain definitional aspects; the murders are repetitive or 'serial' and because 'the murder in itself is the motive', the killings will continue until the killer is apprehended; the majority of serial killers appear to prefer to 'work on their own', which makes them more difficult to trace; there is usually 'little personal connection between the perpetrator and the victim', making the killings often 'classic stranger-perpetuated murders' which are sometimes 'motiveless'; very few serial killers 'display a clearly defined or rational motive'; 'increased spatial mobility and social fragmentation will enable a serial killer to extend and intensify their killing capacity'; and there is often a 'high degree of gratuitous violence' (McLaughlin, ibid.: 404).

Studies suggest that there are four main motivational typologies in relation to serial killers. These are *Visionaries*, which are those killers who claim to be 'directed by "voices" and other egos, where the "instructions" received will justify and legitimate the murders'; *Missionaries*, such as 'clean-up killers' willing to accept responsibility for 'cleansing' societies of 'undesirables', with targets usually chosen from 'deviant' groupings such as sex-workers, who may be able to justify and neutralise their actions 'on the grounds that they are acting on behalf of decent people'; *Hedonists*; for whom pleasure is the 'reward' for murder such as lust killers, fantasy killers and thrill killers; and *Power seekers;* for whom domination 'is the strong motive force', but these killers are 'aware of their behaviour and can describe their motivational state' (McLaughlin, ibid.: 404).

Those such as Canter (1992) have attempted to construct psychological profiles of such serial killers, utilising contextual analysis of victim traits, witness reports, and the method of killing. Emerging from this process will be a detailed list of physical and psychological characteristics, including sex, age, marital status, occupation, 'race', class, sexual preferences, etc., plus a certain amount of intuitive speculation. Canter (ibid.) has suggested that criminal offenders will leave evidence of their personality through their criminal actions. According to Canter, an individual criminal offender's behaviour will exhibit characteristics unique to that person, together with patterns of behaviour that are typical of the subgroup to which they belong. It is the contention here that we can apply a modified notion of Psychobiography to the issue of the analysis of behaviour described by Canter. Canter's reference to the uniqueness of some personal behaviour echoes notions of Psychobiography, of the unique, asocial aspects of the person such as disposition. As McLaughlin (2013: 405) suggests, law enforcement agencies throughout the world 'have realised that they also need to develop sophisticated information collection systems which can coordinate, analyse and review records from different agencies and identify early serial killing patterns'. An acknowledgement that asocial Psychobiography plays a significant role in human behaviour would strengthen any attempts at identification and analysis of such patterns of behaviour, particularly the important idea that Psychobiography embraces the mutuality and plasticity of the relations between genetic and environmental influences. In what follows we examine the application of the meta-concepts of agency–structure, micro–macro and time–space to the study of crime and criminal behaviour.

Agency–structure, micro–macro, time–space

To recap, all these terms are defined at the beginning of the chapter (in the codification of the framework that is entitled 'The Metatheoretical framework'). However, the terms or meta-concepts may be defined briefly as follows. *Agency* is defined in strictly non-reified terms, in which actors are conceptualised as entities capable of formulating and acting upon decisions. *Social structure* is defined as 'social conditions', not a unitary phenomenon but rather contingently reproduced. *Micro–macro* refers to the units/scales of analyses concerned 'with the investigation of varying temporal and spatial extensions of the social' (Sibeon, 2004: 57). *Time–space* refers to relatively neglected concerns within criminology and social theory pertaining to temporality and spatiality.

As previously stated, the emphasis in the 'new' *genetic–social* framework is upon a concept of *agency* which is non-reified. That is to say that it avoids at all cost what Sibeon (2004) identifies as *reification*; the attribution of agency to an entity incapable of formulating and acting upon decisions. However, like Barry Hindess (1986a), Roger Sibeon argues that there are actually *two* main types of actors; *individual human actors* and *social actors* (the latter are termed 'supra-individuals' by Harre, 1981: 150–2), such as organisations, families, committees, central government departments, professional associations, etc.

As Sibeon (2004: 119) puts it, the decisions of such *social actors* 'shape much of the social, economic, and political terrain' in today's society. They are not mere aggregations of the decisions of individual human actors. This relates to the concept of *'emergence'* (ibid.: 120), in the sense of emergent properties arising within social structure. However, the form of *agency* which *social actors* (such as a political party) are capable of, in terms of decisions and actions, is different to that of individual human actors and it is essential not to confuse the two. Thus, as Owen (2009a) suggests in relation to the Human Genome Project, the US Department of Energy (DOE) which originally explored the feasibility of such a mammoth bio-medical undertaking may be regarded as a *'social actor'* or a *'supra-individual'*, depending upon whether we draw upon Hindess (ibid.) and Sibeon (ibid.) or Harre (ibid.) respectively. It cannot, however, be regarded as an individual actor on par with a sentient human being capable of the kind of *agency* human individuals require to formulate/act upon decisions. To imply or suggest so would be to engage in the 'cardinal sin' or *reification*.

In an earlier section of this chapter devoted to the 'cardinal sin' of *reification*, we discussed the theories of *Actor-network* and *posthumanism*, which are greatly at odds with the non-reified conception of agency favoured in the 'new' framework. The point was made that we must reject both kinds of accounts. *Material* and *non-human agency* which attributes causal powers to physical objects (Pickering, 2001) such as stones must be repudiated. Also rejected is Brown's (2013: 488) claim in relation to virtual criminology and cyber-crime to the ends that '[i]n increasing numbers of instances' we are no longer able to distinguish 'human agency, culpability and motivation' from 'technology and non-human objects'. *Contra* Brown, 'the essence of humanity' *is* 'self-evident'. Put simply, it is the human actor who programmes the computer, and only the human actor who is capable of formulating and acting upon decisions. In *genetic–social* theorising, we draw upon the evolutionary psychology of Tooby and Cosmides (1992) in Barkow et al. (1992), which argues for a flexible picture of causality in which genes play a part, but environmental influences and the underlying mechanisms pertaining to psychology play a part also. In a sense, 'nobody is in charge' but, for example, the response to stress in Homo Sapiens depends upon natural selection's intricate, interconnected system in which 'nurture' depends upon genes, and genes require nurture in a form of mutuality and plasticity.

Micro–macro as a meta-concept is applied in the 'new' framework in the sense that Sibeon (2004: 46) intended; in other words, 'micro-macro should not refer to differences in the absolute "size" of social phenomena, but rather, to *relative* differences, in size and to a relational conception of scale associated with the concept of "emergence" '. *Emergence* can be found at 'micro-, mezo- and macro-levels of social process' (ibid.: 76). Sibeon (ibid.) provides an illustratory example of emergence in the case of a 'formal committee or organisation in the public sector' which is 'likely to have institutionalised mechanisms of legitimisation, resource-attraction and decision-making' and 'causal powers' that 'cannot be reduced to the level of its individual members' (Clegg, 1989: 187–8; Holzner, 1978). The point is also made by Sibeon (ibid.), that 'informal groups also have emergent properties' and these can take the form of 'informal systems of rules'.

As Sibeon makes clear, the *spatial* dimension of society has not received even, as much as the (already neglected) *temporal* dimensions. '*Locale*' is posited by Sibeon (ibid.) as a 'bridging concept' to span micro-, mezo- and macro-levels of the social. In his view, 'some local events may be specific to the location in question, others may be connected to the macro-social order' (ibid.). Therefore, we are advised to 'examine the internal properties of local sites as well as the emergent properties of temporal and spatial linkages across sites'. Sibeon (ibid.: 167) stresses the concepts of '*networks*' and '*materials*' in relation to the investigation of the properties of *time–space*.

Networks/systems are 'more or less patterned relations between actors between, for example, positions/roles' and it is partly though not exclusively 'within networks' that materials 'flow' (ibid.). Here the term '*material diffusion*' applies. The concept of materials refers in Sibeon's (2004) work to discourses, social practices and typifications where 'these phenomena are inspected in terms of their tendency to "travel" across spatio-temporal contexts' (ibid.). In terms of explanatory potential with regard to the study of crime and criminal behaviour (and or related areas such as the links between life sciences/'the social'), these concepts can be employed for analytical purposes. We can do so with regard to *materials* and *material diffusion* processes of the kind identified by Sibeon (ibid.). Arguably relevant *materials* to examine in the investigation/study of selected examples of crime and criminal behaviour for analytical purposes might be existing laws. We might regard the *discourse* pertaining to such examples as existing spatially and temporally within and across a variety of *micro* sites characterised by face-to-face interaction, together with *macro* locales such as courts of law, etc. We would then be able to concern ourselves with the detailed analysis of *how* time–space links between relevant (micro, mezo, macro) locales are maintained or changed; the analysis of *micro* (interactional) processes related to the application/handling etc of material *within* the various locales; and the study of actors (individual human actors and or social actors) involved in material diffusion *across* locales across time and space in the social context. The Foucauldian tendency to reify discourse is rejected here. As we have seen earlier, Garland (2001), like Foucault, 'appears to presume that discourses are themselves social actors in the sense of a "crime consciousness" diffused through the media' (Owen, 2012a: 92). Discourses are a form of *material* that require manipulation by human actors. In *genetic–social* theorising, we may conceptualise discourse as a potential influence upon social actors, but to regard discourse as an agent is to engage in illicit reification.

Concluding observations

It is the contention here that a *genetic–social*, metatheoretical framework, which entails a flexible, Realist ontology, relying upon multifactorial analysis rather than reductionist, reified and essentialist analysis, is best equipped to cautiously point a 'way forward' for criminological theory, beyond anti-foundational relativism and the environmentalist rejection of biology. This is an approach which sidesteps the nature versus nurture debate, and which emphasises instead a balanced account of the mutuality between genes and environment. It is the view here that crime may be socially constructed, but there is still a reality 'out there' in which criminologists need to acknowledge that biological factors may 'switch on' genetic impulses to generate behaviour that can be labelled criminal when they interact with other social and psychological factors. However, human beings are reflexive agents with the agency to choose not to engage in criminal activities where they believe that the rewards are outweighed by negative outcomes or actions offend moral prohibitions.

Agency, in turn, is certainly influenced by inherited constitutional variables. These are early days, but hopefully these elements combined make the framework a useful conceptual toolkit. Its methodological generalisations, as opposed to substantive generalisations, its rejection of biophobia and its adoption of a realist social ontology make it a sensitising device with some potential for future theoretical and explanatory use (Owen, 2009a, 2012a).

Bibliography

Albrow, M. (1996) *The Global Age*. Cambridge: Polity Press.
Alexander, B. (2003) *Rapture: How Biotechnology Became the New Religion*. New York: Basic Books.
Alexander, J. (1992) 'General theory in the postpositivistic mode: The "epistemological dilemma" and the search for present reason', in S. Seidman and D.G. Wagner (eds), *Postmodernism and Social Theory*. Cambridge, MA and Oxford: Blackwell.
Anderson, B. (1991) *Imagined Communities: Reflections on the Origins and Spread of Nationalism*. London: Verso.
Archer, M. (1982) 'Structuration versus morphogenesis: On combining structure and action', *British Journal of Sociology*, 33(4): 445–83.
Archer, M. (1988) *Culture and Agency: The Place of Culture in Social Theory*. Cambridge: Cambridge University Press.
Archer, M. (1995) *Realist Social Theory: The Morphogenetic Approach*. Cambridge: Cambridge University Press.
Archer, M. (1996) 'Social integration and system integration: Developing the distinction', *Sociology*, 30(4): 679–99.
Archer, M. (1998) 'Social theory and the analysis of society', in T. May and M. Williams (eds), *Knowing the Social World*. Buckingham: Open University Press.
Archer, M. (2000) *Being Human: The Problem of Agency*. Cambridge: Cambridge University Press.
Armstrong, D. (1987) 'Theoretical terrains in biopsychosocial medicine', *Social Science and Medicine*, 25: 1213–18.
Association of British Insurers. (1997) *Life Insurance and Genetics: A Policy Statement*.
Barkow, J., Cosmides, L. and Tooby, J. (1992) *The Adapted Mind*. Oxford: Oxford University Press.
Barnes, B. and Shapin, S. (eds) (1979) *Natural Order: Historical Studies of Scientific Culture*. London: Sage.
Barrett, L., Dunbar, R.I.M. and Lycett, J. (2002) *Human Evolutionary Psychology*. London: Belgrave.
Barthes, R. (1967) *Elements of Semiology*. London: Cape.
Baudrillard, J. (1983) *In the Shadow of the Silent Majorities or the End of the Social and Other Essays*. New York: Semiotext(e).
Bauman, Z. (1992) *Intimations of Postmodernity*. London: Routledge.
Beatty, J. (2000) 'Origins of the U.S. human genome project: The changing relationship between genetics and national security', in P.R. Sloan (ed) *Controlling Our Destinies: Historical, Philosophical, Ethical, and Theological Perspectives on the Human Genome Project*. Notre Dame: Notre Dame Press.

Beck, U. (1992) *The Risk Society*. London: Sage.
Beck, U. (1999) *World Risk Society*. Cambridge: Polity Press.
Beck, U. (2002) 'The silence of words and the political dynamics in the world risk society', *Logos*, 1(4): 1–18.
Becker, J.B., Breedlove, S.M., and Crews, D. (1992) *Behavioural Endocrinology*. Cambridge, MA: MIT Press.
Bell, D. (1973) The Coming of Post-Industrial Society. London: Heinemann.
Benjamin, J., Li, L., Patterson, C. and Greenberg, B.D. (1996) 'Population and familial association between the D4 dopamine receptor gene and measures of novelty seeking', *Nature*, 12: 81–4.
Benton, T. (1991) 'Biology and social science: Why the return of the repressed should be given a (cautious) welcome', *Sociology*, 25(1): 1–29.
Benton, T. (1993) Natural Relations: Ecology, Animal Rights and Social Justice. London: Verso.
Benton, T. (1994) 'Biology and social theory in the environmental debate', in M. Redclift and T. Benton (eds), *Social Theory and the Global Environment*. London: Routledge.
Benton, T. (1998) 'Realism and social science: Some comments on Roy Bhaskar's "The Possibility of Naturalism"', in M. Archer, R. Bhaskar, A. Collier, T. Lawton and A. Norrie (eds), *Critical Realism: Essential Readings*. London: Routledge.
Benton, T. (1999) 'Evolutionary psychology: A new paradigm or just the same old reductionism?', *Advances in Human Ecology*, 8: 65–98.
Benton, T. (2003) 'Ecology, health and society: A red-green perspective', in S.J. Williams, L. Birke and G. Bendelow (eds), *Debating Biology: Sociological Reflections on Health, Medicine and Society*. London: Routledge.
Benton, T. and Craib, I. (2001) *The Philosophy of Science*. London: Palgrave.
Bennington, J. and Harvey, J. (1994) 'Spheres or tiers? The significance of transnational local authority networks', *Local Government Policy Making*, 20(5): 21–30.
Bennington, J. and Taylor, M. (1993) 'Changes and challenges facing the UK welfare state in the EU of the 1990s', *Policy and Politics*, 21(2): 121–34.
Berger, P. and Luckmann, T. (1971) *The Social Construction of Reality*. Harmondsworth: Penguin.
Best, S. and Kellner, D. (1991) *Postmodern Theory: Critical Interrogations*. London: Macmillan.
Betts, K. (1986) 'The conditions of action, power and the problem of interests', *Sociological Review*, 34(1): 39–64.
Bhaskar, R. (1989a) Reclaiming Reality: A Critical Introduction to Contemporary Philosophy. London: Verso.
Bhaskar, R. (1989b) *The Possibility of Naturalism*. London: Harvester Wheatsheaf.
Biggs, S. (1999) *The Mature Imagination*. Milton Keynes: OUP.
Biggs, S. and Powell, J.L. (1999) 'Surveillance and elder abuse: The rationalities and technologies of community care', *Journal of Contemporary Health*, 4(1): 43–9.

Bijker, W. (1995) *Of Bicycles, Bakelites and Bulbs*. New York: Routledge.
Birke, L. (2003) 'Shaping biology: Feminism and the idea of "the biological"', in S.J. Williams, L. Birke and G. Bendelow (eds), *Debating Biology: Sociological Reflections on Health, Medicine and Society*. London: Routledge.
Blackledge, D. and Hunt, B. (1993) *Sociological Interpretations of Education*. London: Routledge.
Bloor, D. (1976) *Knowledge and Social Imagery*. London: Routledge.
Bodmer, W.F. and McKie, R. (1994) *The Book of Man*. London: Little Brown.
Bogaert, A.F. and Fisher, W.A. (1995) 'Predictors of university men's number of sexual partners', *The Journal of Sex Research*, 32: 119–30.
Bojorquez, J. (1994) 'In his image', *Sacramento Bee*, 18.
Boseley, S. (2005) 'Drive launched to find egg and sperm donors', *Guardian*, Wed 26 January.
Bourdieu, P. (1977) *Outline of a Theory of Practice*. Cambridge: Cambridge University Press.
Bourdieu, P. (1984) *Distinction: A Social Critique of the Judgement of Taste*. Cambridge, MA: Harvard University Press.
Bourdieu, P. (1990) In Other Words: Essays Towards a Reflexive Sociology. Cambridge: Polity Press.
Bourdieu, P. (1998) *Practical Reason: On the Theory of Action*. Cambridge: Polity Press.
Brennan, Z. and Syal, R. (17 August 1997) 'Sperm donor payments to end', *Sunday Times* (London), 7.
Bretherton, C. (1996) 'Introduction: Global politics in the 1990s', in C. Bretherton and G. Ponton (eds), *Global Politics: An Introduction*. Oxford: Blackwell.
Bridenthall, R., Grossman, A. and Kaplan, M. (eds) (1984) *When Biology Became Destiny: Women in Weimar and Nazi Germany*. New York: Monthly Review Press.
Brittan, S. (1998) 'Essays, moral, political and economic', *Hume Papers on Public Policy*, vol. 6. Edinburgh: Edinburgh University Press.
Brodwin, P.E. (ed) (2001) *Biotechnology and Culture: Bodies, Anxieties, Ethics*. Indiana: Indiana University Press.
Bromley, D.W. and Segerson, K. (eds) (1992) The Social Response to Environmental Risk: Policy Formation in an Age of Uncertainty. London: Springer.
Brown, K. (2000) 'The human genome business today', *Scientific American*, 283(1): 7.
Brown, S. (2009) 'Virtual criminology', in E. McLaughlin and E. Muncie (eds) *The Sage Dictionary of Criminology* (2nd Edition). London: Sage Publishers.
Brown, S. (2013) 'Virtual criminology', in E. McLaughlin and J. Muncie (eds), *The Sage Dictionary of Criminology* (3rd Edition). London: Sage.
Brown, R.S. and Marshall, K. (eds) (1993) *Advances in Genetic Information: A Guide for State Policy Makers*. Lexington, KY: Council of State Governments.

Brunner, H.G., Nelen, M.R. and Van Zandvoort, P. (1993a) 'X-linked borderline mental retardation with prominent behavioural disturbance: Phenotype, genetic localization and evidence for disturbed monoamine metabolism', *American Journal of Human Genetics*, 52: 1032.

Brunner, H.G., Nelen, M.R., Breakefield, X.O. and Ropers, H.H. (1993b) 'Abnormal behaviour associated with a point mutation in the structural gene for monoamine oxidase A', *Science*, 262: 578–80.

Bucchi, M. (2004) Science in Society: An Introduction to Social Studies of Science. London and New York: Routledge.

Buctuanon, E.M. (2001) 'Globalisation of biotechnology: The agglomeration of dispersed knowledge and information and its implications for the political economy of technology in developing countries', *New Genetics and Society*, 20(1): 25–41.

Bulmer, M. (1985) 'The rejuvenation of community studies? Neighbours, networks and policy', *The Sociological Review*, 33(3): 430–48.

Bulmer, M. (1986) *Neighbours: The Works of Philip Abrams*. Cambridge: Cambridge University Press.

Bury, M.R. (1995) 'The body in question', *Medical Sociology News*, 21(1): 36–48.

Bury, M.R. (1997) Health and Illness in a Changing Society. London: Routledge.

Buss, D.M. (1999) *Evolutionary Psychology*. Boston: Allyn and Bacon.

Buss, D. (2000) 'The evolution of happiness', *American Psychologist*, 55 (1): 15–23.

Caldwell, J.H. (1984) 'Babies by scientific selection', *Scientific American*, 150: 124–25.

Calhoun, C., Gerteis, J., Moody, J., Pfaff, S. and Virk, I. (eds) (2002) *Contemporary Sociological Theory*. Oxford: Blackwell.

Callon, M. (1986) 'Some elements of a sociology of translation: Domestication of the scallops and the fisherman of St. Brieuc Bay', in J. Law (ed), *Power Action and Belief: A New Sociology of Knowledge?* London: Routledge.

Callon, M. (1991) 'Techno-economic networks and irreversability', in J. Law (ed) *Power.Action and Belief: A New Sociology of Knowledge*. London: Routledge.

Callon, M. and Latour, B. (1981) 'Unscrewing the big Leviathan: How actors macro-structure reality and how sociologists help them to do so', in K. Knorr-Cetina and A.V. Cicourel (eds), *Advances in Social Theory and Methodology: Towards an Integration of Micro-and Macro-Sociologies*. London: Routledge.

Cannon, W.B. (1914) 'The interrelations of emotion as suggested by recent physiological researchers', *American Journal of Psychology*, 25: 256–82.

Cantor, C. (1992) 'The challenges to technology and information', in D.J. Kevles and L. Hood (eds), *The Code of Codes*. Cambridge, MA: Harvard University Press.

Canter, D. (1992) *Criminal Psychology*. London; Routledge.

Cantor, J.M., Blanchard, R., Paterson, A.D. and Bogaert, A.F. (2002) 'How many gay men owe their sexual orientation to fraternal birth order?', *Archives of Sexual Behaviour*, 31: 63–71.

Caplan, A.L. (2000) 'What's morally wrong with Eugenics?', in P.R. Sloan (ed), Controlling our Destinies: Historical, Philosophical, Ethical, and Theological Perspectives on the Human Genome Project. Notre Dame: Notre Dame Press.
Cavadino, M. and Dignan, J. (2006) *Penal Systems: A Comparative Approach*. London: Sage.
Cazillis, M. (2001) 'Facing omnipresent ethical problems', *EUROPA, European Commission Research Information Centre*, update 1 December 2001.
Churchland, S. and Grush, R. (1999) 'Computation and the brain', in R.A. Wilson and F.C. Kiel (eds), *The MIT Encyclopedia of the Cognitive Sciences*. Cambridge, MA: MIT Press.
Clarke, J. (2013) 'Birmingham "School"', in E. McLaughlin and J. Muncie (eds), *The Sage Dictionary of Criminology* (3rd Edition). London: Sage.
Clegg, S. (1989) *Frameworks of Power*. London: Sage.
Coburn, D. (2000a) 'Income inequality, social cohesion and the health status of populations: The role of neoliberalism', *Social Science and Medicine*, 51: 135–46.
Coburn, D. (2000b) 'A brief response', *Social Science and Medicine*, 51: 1009–10.
Cohen, C. (1996) *New Ways of Making Babies: The Case of Egg Donation*. Bloomington: Indiana University Press.
Cohen, I. (1987) 'Structuration theory and social praxis', in A. Giddens and J. Turner (eds), *Social Theory Today*. Cambridge: Polity Press.
Cohen, I. (1989) Structuration Theory: Anthony Giddens and the Constitution of Social Life. London: Macmillan.
Coleman, R. and Sim. J. (2013) 'Left idealism', in E. McLaughlin and J. Muncie (eds), *The Sage Dictionary of Criminology* (3rd Edition). London: Sage.
Collier, R. (1998) Masculinities, Crime and Criminology. London: Sage.
Collins, H. and Kusch, M. (1998) *The Shape of Actions: What Humans and Machines Can Do*. Cambridge, MA: MIT Press.
Collins, H.M. and Pinch, T. (1993) *The Golem: What Everyone Should Know About Technology*. Cambridge: Cambridge University Press.
Connell, R. (1995) *Masculinities*. Cambridge: Polity Press.
Connell, R. (2000) *The Men and the Boys*. Berkeley: University of California Press.
Cosmides, L. and Tooby, J. (1992) 'Cognitive adaptations for social exchange', in J. Barkow, L. Cosmides and J. Tooby (eds), *The Adapted Mind: Evolutionary Psychology and the Generation of Culture*. New York: Oxford University Press.
Cosmides, L. and Tooby, J. (1997) *Evolutionary Psychology: A Primer*. Available at http://www.psych.ucsb.edu/research/cep/primer.html.
Crawford, C.B. and Anderson, J.L. (1989) 'Sociobiology: An environmentalist discipline?', *American Psychologist*, 44(12): 1449–59.
Curry, O. (2003) 'Evolutionary psychology: "fashionable ideology" or "new foundation"?', *Human Nature Review*, 3(28 January): 81–92.
Daly, M. and Wilson, M. (1988) Homicide. New York: Aldine de Gruyter.
Daly, M. and Wilson, M. (1998) *The Truth About Cinderella: A Darwinian View of Parental Love*. London: Weidenfeld and Nicolson.

Daly, M. and Wilson, M. (2001) 'An assessment of some proposed exceptions to the phenomenon of nepotistic discrimination against stepchildren', *Annales Zoologici Fennici*, 38: 287–96.
Danaher, G., Schirato, T. and Webb, J. (2000) *Understanding Foucault*. London: Sage.
David, M. (2002) 'The sociological critique of evolutionary psychology: Beyond mass modularity', *New Genetics and Society*, 21(3): 303–13.
David, M. and Kirkhope, J. (2005) 'Cloning/stem cells and the meaning of life', *Current Sociology*, 53(2): 367–81.
Davies, I. (1993) 'Cultural theory in Britain: Narrative and episteme', *Theory, Culture and Society*, 10(3): 115–54.
Dawkins, R. (1976) *The Selfish Gene*. Oxford: Oxford University Press.
Dawkins, R. (1982) *The Extended Phenotype*. London: W.H.Freeman.
Dawkins, R. (1986) *The Blind Watchmaker*. New York: W.W. Norton and Co.
Dawkins, R. (1992) 'Progress', in E. Fox Keller and E. Lloyd (eds), *Keywords in Evolutionary Psychology*. Cambridge, MA: Harvard University Press.
Degener, T. (1990) 'Female self-determination between feminist claims and "Voluntary" eugenics, between "Rights" and ethics, issues in reproductive and genetic engineering', *Journal of International Feminist Analysis*, 3(2): 87–99.
Delanty, G. (1999) *Social Theory in a Changing World: Conceptions of Modernity*. Cambridge: Polity Press.
Delanty, G. (2002) 'Constructivism, sociology and the new genetics', *New Genetics and Society*, 21(3): 279–89.
Dennis, C. and Gallagher, R. (eds) (2001) *The Human Genome*. Hampshire and New York: Palgrave.
Derbyshire, D. and Highfield, R. (2003) 'Women better than men at instant maths', *The Daily Telegraph*, 11 September.
Derrida, J. (1982) 'Différence', in trans. A. Bass, *Margins of Philosophy*. Hemel Hempstead: Harvester Wheatsheaf.
Dickens, P. (1996) *Reconstructing Nature: Alienation, Emancipation and the Division of Labour*. London: Routledge.
Dickens, P. (2000) *Social Darwinism*. Buckingham: Open University Press.
Dietz, T.R., Frey, R.S. and Rosa, E. (1997) 'Risk, technology and society', in R.E. Dunlap and W. Michelson (eds), *Handbook of Environmental Sociology*. Westport, CT: Greenwood Press.
Dobash, R.E. and Dobash, R.P. (1992) *Rethinking Violence Against Women*. London: Sage.
Dominelli, L. (1997) *Sociology for Social Work*. London: Macmillan.
Dreger, A.D. (2000) 'Metaphors of morality in the human genome project', in P.R. Sloan (ed), *Controlling Our Destinies: Historical, Philosophical, Ethical, and Theological Perspectives on the Human Genome Project*. Notre Dame: Notre Dame Press.
Dunbar, R.I.M. (1996) *Gossip, Grooming and the Evolution of Language*. London: Faber and Faber.

Dunbar, R.I.M. (2003a) 'Why are apes so smart?', in P. Keppeler and M. Pereira (eds), *Primate Life Histories and Socioecology*. Chicago: Chicago University Press.
Dunbar, R.I.M. (2003b) 'The social brain: Mind, language and society in evolutionary perspective', *Annual Review of Anthropology*, 32: 163–81.
Dunbar, R.I.M. (2004) *The Human Story*. London: Faber and Faber.
Dunbar, R.I.M. (2006) 'Taking social intelligence seriously', in R.A. Peel and M. Zekin (eds), *Human Ability: Genetic and Environmental Influences*, proceedings of a conference organised in London: The Galton Institute.
Dunbar, R.I.M, Knight, C. and Power, C. (1999) *The Evolution of Culture*. Edinburgh: Edinburgh University Press.
Dupre, J. (2001) *Human Nature and the Limits of Science*. Oxford: Oxford University Press.
Durant, J., Baur, M.W. and Gaskell, G. (1998) *Biotechnology in the Public Sphere: A European Sourcebook*. London: NMS Trading.
Durkheim, E. (1982) *The Rules of Sociological Method*. London: Macmillan.
Duster, T. (1990) *Backdoor to Eugenics*. New York: Routledge.
Dye, T. (1998) *Understanding Public Policy*. Upper Saddle River, NJ: Prentice Hall.
Elias, N. (1978) *What is Sociology?* London: Hutchison.
Elliot, P. and Mandell, N. (1995) 'Feminist theories', in N. Mandell (ed), *In Answering Back: Girls, Boys and Feminism in Schools*. Scarborough, ON: Prentice Hall Canada Inc.
Elstad, J.I. (1998) 'The psycho-social perspective on social inequalities in health', *Sociology of Health and Illness*, 20(5): 598–618.
Eltzinga, A. and Jamison, A. (1995) 'Changing policy agendas in Science and Technology', in S. Jasonoff (ed), *Handbook of Science and Technology Studies*. Thousand Oaks: Sage.
Enard, W., Przeworski, M., Fisher, S.E., Lai, C.S.L., Wiebe, V., Kitano, T., Monaco, A.P. and Paabo, S. (2002) 'Molecular evolution of FOXP2, a gene involved in speech and language', *Nature*, 418: 869–72.
Ericson, R.V. and Doyle, A. (eds) (2003) *Risk and Morality*. Toronto: University of Toronto Press.
Ettorre, E. (2000) 'Reproductive genetics, gender and the body: 'Please doctor, may I have a normal baby?', *Sociology*, 34(3): 403–20.
Etzkowitz, H. and Webster, A. (1995) 'Science as intellectual property', in S. Jasonoff (ed.), *Handbook of Science and Technology Studies*. Thousand Oaks: Sage.
European Parliament. (1990) Ethical and Legal Problems of Genetic Engineering and Human Artificial Insemination. Strasbourg.
European Parliament. (11 March 1997). 'Resolution on cloning', *Proceedings of the European Parliament*. Strasbourg.
Ewing, C.M. (1998) 'Tailored genes: IVF, Genetic Engineering and Eugenics, Reproductive and Genetic Engineering', *Journal of International Feminist Analysis*, 3(2): 119–23.

Ezzell, C. (1987) 'First ever animal patent issues in United States', *Nature*, 332: 21.
Farganis, S. (1994) *Situating Feminism: From Thought to Action*. Thousand Oaks, CA, and London: Sage.
Featherstone, M. (1988) 'In pursuit of the postmodern', *Theory, Culture and Society*, 5: 195–215.
Featherstone, M. and Wernick, A. (1995) *Images of Ageing*. London: Routledge.
Feingold, W. (1976) *Artificial Insemination*. New York: Academic Press.
Ferrell, J. (2012) 'Outline of a criminology of drift', in S.Hall and S.Winlow (eds), *New Directions in Criminological Theory*. London: Routledge.
Filley, C.M., Price, B.H., Nell, V., Antoinette, T., Morgan, A.S., Bresnahan, J.K., Pincus, J.H., Gelbort, M.M., Weissberg, M. and Kelly, J.P. (2001) 'Towards an understanding of violence: Neurobehavioural aspects of unwarranted physical aggression: Aspen neurobehavioural conference consensus statement', *Neuropsychiatry, Neuropsychology and Behavioural Neurology*, 14: 1–14.
Fleising, U. (2001) 'Genetic essentialism, *mana*, and the meaning of DNA', *New Genetics and Society*, 20(1): 43–57.
Fodor, J. (1983) *The Modularity of Mind*. Cambridge, MA: MIT Press.
Fodor, J. (2000) *In Critical Condition*. Cambridge, MA: MIT Press.
Fodor, J. (2001) *The Mind Doesn't Work That Way*. Cambridge, MA: MIT Press.
Foucault, M. (1970) The Order of Things: An Archaeology of the Human Sciences. London: Tavistock.
Foucault, M. (1972) *The Archaeology of Knowledge*. New York: Pantheon Books.
Foucault, M. (1980a) *Power/Knowledge*. New York: Pantheon Books.
Foucault, M. (1980b) *The History of Sexuality*. New York: Vintage Books.
Foucault, M. (1982) 'The subject and power', in H.L. Dreyfus and P. Rabinow (eds), *Michael Foucault: Beyond Structuralism and Hermeneutics, with an Afterword by Michel Foucault*. Brighton: Harvester Wheatsheaf.
Foucault, M. (1991) 'Governmentality', in G. Burchell (ed), *The Foucault Effect: Studies in Governmentality*. Hemel Hempstead: Harvester Wheatsheaf.
Fox, C.J. and Miller, H.T. (1995) *Postmodern Public Administration*. London: Sage.
Fox, N.J. (1991) 'Postmodernism, rationality and the evaluation of health care', *Sociological Review*, 39(4): 709–44.
Frank, A.W. (1990) 'Bringing bodies back in: A decade review', *Theory, Culture and Society*, 7(1): 131–62.
Freese, J., Li, J.C.A. and Wade, L.D. (2003) 'The potential relevances of biology to social inquiry', *Annual Review of Sociology*, 29: 233–56.
Freund, P.E.S. (1988) 'Bringing society into the body', *Theory and Society*, 17: 839–64.
Freund, P.E.S. (1990) 'The expressive body: A common ground for the sociology of emotions and health and illness', *Sociology of Health and Illness*, 12(4): 452–77.
Freund, P.E.S. (1998) 'Social performances and their discontents: The biopsychosocial aspects of dramaturgical stress', in G. Bendelow and S.J. Williams

(eds), *Emotions in Social Life: Critical Themes, Contemporary Issues*. London: Routledge.

Freund, P.E.S. (2001) 'Bodies, disability and spaces: The social model of disability and disabling spatial organisations', *Disability and Society*, 16, 5, 689–706.

Friedrichs, R.W. (1972) *A Sociology of Sociology*. New York: Free Press.

Gagnon, J.H. and Simon, W. (1973) *Sexual Conduct*. London: Hutchinson.

Galton Institute. (1999) 'Notes of the quarter', *The Galton Institute Newsletter*, 2: 8.

Garland, D. (2001) *The Culture of Control: Crime and Social Order in Contemporary Society*. Oxford: Oxford University Press.

Gellner, E. (1993) *Postmodernism, Reason and Religion*. London and New York: Routledge.

Gibbons, M., Limoges, C., Nowotny, H., Schwartzman, S., Scott, P. and Trow, M. (1994) The New Production of Knowledge: The Dynamics of Science and Research in Contemporary Societies. London: Sage.

Gibbs, J.P. (1989) *Control: Sociology's Central Notion*. Urbana, IL: University of Illinois Press.

Giddens, A. (1979) Central Problems in Social Theory: Action, Structure and Contradiction in Social Analysis. Berkeley, CA: University of California Press.

Giddens, A. (1981) 'Agency, institution and time-space analysis', in K. Knorr-Cetina and A.V. Cicourel (eds), *Advances in Social Theory and Methodology: Towards an Integration of Micro-and Macro-Sociologies*. London: Routledge.

Giddens, A. (1982) Profiles and Critiques in Social Theory. London: Macmillan.

Giddens, A. (1984) *The Constitution of Society*. Cambridge: Polity Press.

Giddens, A. (1987) *Social Theory and Modern Sociology*. Cambridge: Polity Press.

Giddens, A. (1989) 'A reply to my critics', in D. Held and J. Thompson (eds), *Social Theory and Modern Societies: Anthony Giddens and his Critics*. Cambridge: Cambridge University Press.

Giddens, A. (1991a) 'Structuration theory: Past, present and future', in C. Bryant and D. Jary (eds), *Giddens's Theory of Structuration: A Critical Appreciation*. London: Routledge.

Giddens, A. (1991b) *Modernity and Self-Identity*. Cambridge: Polity Press.

Giddens, A. (1993) *Sociology*. Cambridge: Polity Press.

Gigerenzer, G., Todd, P. and the ABC Research Group. (1999) *Simple Heuristics that Make US Smart*. Milton Keynes: Open University Press.

Goffman, E. (1983) 'The interaction order', *American Sociological Review*, 48(3): 1–17.

Gottweis, H. (2005) 'Governing genomics in the 21st century: Between risk and uncertainty', *New Genetics and Society*, 24(2): 175–94.

Grabosky, P.N. (2001) 'Virtual criminality: Old wine in new bottles?', *Social and Legal Studies*, 10: 243–9.

Gurnah, A. and Scott, A. (1992) The Uncertain Science: Criticisms of Sociological Formalism. London: Routledge.
Habermas, J. (1986) The Theory of Communicative Action. Vol. 1: Reason and the Rationalization of Society. Cambridge: Polity Press.
Habermas, J. (1987) The Theory of Communicative Action. Vol. II: The Critique of Functionalist Reason. Cambridge: Polity Press.
Habermas, J. (1989) The Structural Transformation of the Public Sphere. Cambridge, MA: MIT Press.
Hall, S. (1985) 'Authoritarian populism: A reply to Jessop et al.', *New Left Review*, 151: 115–24.
Hall, S. (2008) 'Daubing the drudge of fury: Men, violence and the piety of the hegemonic masculinity thesis', in S. Tomsen (ed), *Men, Masculinities and Crime*. International Library of Criminology, Criminal Justice and Penology, vol. 5. Aldershot: Ashgate.
Hall, S., Critcher, C., Jefferson, T., Clarke, J. and Roberts, B. (1978) *Policing the Crisis: Mugging, the State and Law and Order*. London: Macmillan.
Hall, S.S. (2003) Merchants of Immortality: Chasing the Dream of Human Life Extension. New York: Houghton Miffin.
Hall, S. and Winlow, S. (eds) (2012) *New Directions in Criminological Theory*. London: Routledge.
Hamer, D. and Copeland, P. (1999) Living With Our Genes: Why They Matter More Than You Think. London: Macmillan.
Hamerton, J.L., Evans, J.A. and Stranc, L. (1993) 'Prenatal diagnosis in Canada', *Research Volumes of the Royal Commission on New Reproductive Technologies*. Ottawa: H.M.S.O.
Hanmer, J. (1993) 'Reproductive technology: The future for women', in J. Rothchild (ed), *Machina Ex Dea: Feminist Perspectives on Technology*. New York: Pergamon.
Hannigan, J.A. (1995) Environmental Sociology: A Social Constructionist Perspective. London: Routledge.
Haraway, D. (1997) Modest_Witness @ Second Millenium. Female Man Meets Oncomouse™. New York: Routledge.
Harre, R. (1981) 'Philosophical aspects of the macro-micro problem', in K.C. Knorr-Cetina and A.V. Cicourel (eds), *Advances in Social Theory and Methodology: Towards an Integration of Micro-And Macro-Sociologies*. London: Routledge.
Harris, J.R. (1998) The Nurture Assumption. London: Bloomsbury.
Harvey, D. (1989) *The Condition of Postmodernity*. Oxford: Blackwell.
Hay, C. (2002) Political Analysis: A Critical Introduction. Basingstoke: Palgrave.
Hayek, F.A. (1960) *The Constitution of Liberty*. London: Routledge and Kegan Paul.
Herrnstein, R.J. and Murray, C. (1994) *The Bell Curve*. New York: Basic Books.
Hester, M., Kelly, L. and Radford, J. (1996) *Women, Violence and Male Power*. Buckingham: Open University Press.

Hindess, B. (1986a) 'Actors and social relations', in M.L. Wardell and S.P. Turner (eds), *Sociological Theory in Transition*. London: Allen and Unwin.
Hindess, B. (1986b) 'Interests in political analysis', in J. Law (ed), *Power Action and Belief: A New Sociology of Knowledge?* London: Routledge.
Hindess, B. (1988) *Choice. Rationality and Social Theory*. London: Unwin Hyman.
Hochschild, A.R. (1983) *The Managed Heart: Commercialisation and Human Feeling*. Berkeley: University of California Press.
Hochschild, A.R. (1990) The Second Shift: Working Parents and the Revolution at Home. London: Piatkus.
Holmwood, J. (1996) Founding Sociology? Talcott Parsons and the Idea of General Theory. London and New York: Longman.
Hopkins Burke, R. (2009) *An Introduction to Criminological Theory* (3rd Edition). Oxon: Willan Publishing.
Holzner, B. (1978) 'The construction of social actors: An essay on social identities', in T. Luckmann (ed), *Phenomenology and Sociology*. Harmondsworth: Penguin.
Irwin, A. (2001) *Sociology and the Environment*. Cambridge: Polity Press.
Irwin, A. and Lynne, B. (eds) (2003) *Misunderstanding Science?: The Public Reconstruction of Science and Technology*. Cambridge: Cambridge University Press.
James, S. (1997) British Government: A Reader in Policy Making. London: Routledge.
James, W. (1890) *Principles of Psychology*. New York: Holt.
Jameson, F. (1991) Postmodernism or the Cultural Logic of Late Capitalism. London: Verso.
Jefferson, T. (2007) 'Masculinities', in E. McLaughlin and J. Muncie (eds), *The Sage Dictionary of Criminology* (2nd Edition). London: Sage.
Jones, M. (1996) 'Posthuman agency: Between theoretical traditions', *Sociological Theory*, 14(3): 290–309.
Jones, M. and Salter, B. (2004) 'The governance of human genetics: Policy discourse and constructions of public trust', *New Genetics and Society*, 22(1): 21–41.
Kandel, E., Schwartz, J. and Jessel, T. (1995) *Principles of Neural Science*. New York: Elsevier.
Karmiloff-Smith, A. (1992) Beyond Modularity: A Developmental Perspective on Cognitive Science. Cambridge, MA: MIT Press.
Karmiloff-Smith, A. (2000) 'Why babies' brains are not Swiss Army knives', in H. Rose and S. Rose (eds), *Alas Poor Darwin: Arguments against Evolutionary Psychology*. London: Jonathan Cape.
Katz, S. (1997) *Disciplining Old Age*. Charlottesville: University of Virginia.
Kaupen-Haas, H. (1988) 'Experimental obstetrics and national socialism: The conceptual basis of reproductive technology today', *Reproductive and Genetic Engineering: Journal of International Feminist Analysis*, 1(2): 127–32.
Kegan, P. (1991) *Knowledge and Social Imagery* (2nd Edition). Chicago: Chicago University Press.

Keller, E.F. (1995) Refiguring Life: Metaphors of Twentieth-Century Biology. New York: Columbia Press.

Kelsoe, J., Ginns, E., Egeland, J., Gerhard, D.S., Goldstein, A.M., Bale, S.J., Pauls, D.L., Long, R.T., Kidd, K.K., Conte, G., Housman, E., and Paul, S.M. (1989) 'Reevaluation of the linkage relationship between chromosome 11p loci and the gene for bipolar affective disorder in the old order amish', Nature, 342: 238–43.

Kennedy, J., Giuffra, L., Moises, H., et al. (1988) 'Evidence against linkage of schizophrenia to markers on chromosome 5 in a normal Swedish pedigree', Nature, 336: 16–69.

Kerr, A. (2003) 'Rights and responsibilities in the new genetics era', Critical Social Policy, 23(2): 208–26.

Kevles, D.J. (1985) In the Name of Eugenics. New York: Knopf.

Kevles, D.J. and Hood, L. (1992) 'Reflections', in D.J. Kevles and L. Hood (eds), The Code of Codes. Cambridge, MA: Harvard University Press.

King, T.J. and Briggs, R. (1956) 'Serial transplantation of embryonic nuclei', Cold Spring Harbor Symposium on Quantitative Biology, 21: 271–90.

Kitcher, P. (1996) The Lives to Come. New York: Simon and Schuster.

Klinteberg, B. (1996) 'Biology, norms, and personality: A developmental perspective', Neuropsychobiology, 34: 146–54.

Knorr-Cetina, K. (1981) The Manufacture of Knowledge: An Essay on the Constuctivist and Contextual Nature of Science. Oxford: Pergamon.

Knorr-Cetina, K.D. and Cicourel, A.V. (eds) (1981) Advances in Social Theory and Methodology: Towards an Integration of Micro- and Macro-Sociologies. London: Routledge

Kolata, G. (1997) Clone. London: Alton Lane.

Kollek, R. (1990) 'The limits of experimental knowledge: A feminist perspective on the ecological risks of Genetic Engineering', Issues in Reproductive and Genetic Engineering. Journal of International Feminist Analysis, 3(2): 125–35.

Kooiman, J. (2003) Governing as Governance. London: Sage.

Koshland, D.E. (1988) 'The future of biological research: What is possible and what is ethical?', Science, 3, 11–15.

Koski, C.A. (2005) 'The human genome project: An examination of its challenge to the technological imperative', New Genetics and Society, 24(3): 267–81.

Kuhn, T.S. (1962) The Structure of Scientific Revolutions (2nd Edition). Chicago: Chicago University Press.

Kumar, K. (1995) From Post-Industrial to Post-Modern Society: New Theories of The Contemporary World. Oxford: Blackwell.

Kupferman, I. (1992) 'Genetic determinants of behaviour', in E. Kendel, J. Schwartz and T. Jessel (eds), Principles of Neural Science. New York: Elsevier.

Kurzban, R. (2002) 'Alas poor evolutionary Psychology: Unfairly accused, unjustly condemned', Human Nature Review, 2: 99–109. Available at http://human-nature.com/nibbs/02/apd.htm.

Lacan, J. (1977) Ecrits: A Selection. London: Tavistock.

Laclau, E. and Mouffe, C. (1985) *Hegemony and Socialist Strategy*. London: Verso.
Laclau, E. and Mouffe, C. (1987) 'Post-marxism without apologies', *New Left Review*, 166: 79–106.
Lai, C.S., Fisher, S.E., Hurst, J.A., Vargha-Khadem, F. and Monaco, A.P. (2001) 'A forkhead-domain gene is mutated in a severe speech and language disorder', *Nature*, 413: 519–23.
Latour, B. (1986) 'The powers of association', in J. Law (ed), *Power Action and Belief: A New Sociology of Knowledge?* London: Routledge.
Latour, B. (1987) *Science in Action*. Cambridge, MA: Harvard University Press.
Latour, B. (1988) *The Pasteurization of France*. Cambridge, MA: Harvard University Press.
Latour, B. (1993) *We Have Never Been Modern*. London: Harvester Wheatsheaf.
Latour, B. and Woolgar, S. (1979) *Laboratory Life: The Social Construction of Scientific Facts*. Princeton: Princeton University Press.
Lau, S. (1995) 'Report from Hong Kong', *Cambridge Quarterly of Healthcare Ethics*, 4: 364–66.
Law, J. (1986a) 'On power and its tactics: A view from the sociology of science', *Sociological Review*, 34(1): 1–38.
Law, J. (1986b) Power Action and Belief: A New Sociology of Knowledge? London: Routledge.
Law, J. (ed) (1991a) A Sociology of Monsters: Essays on Power, Technology and Domination. London: Routledge.
Law, J. (1991b) 'Power, discretion and strategy', in J. Law (ed), *A Sociology of Monsters: Essays on Power, Technology and Domination*. London: Routledge.
Law, J. (1994) *Organising Modernity*. Oxford: Blackwell.
Law, J. (ed) (1999) *Actor Network Theory and After*. London: Blackwell.
Layder, D. (1984) 'Sources and levels of commitment in actors' careers', *Work and Occupations*, 111(2): 198–216.
Layder, D. (1993) *New Strategies in Social Research: An Introduction and Guide*. Oxford: Polity Press.
Layder, D. (1994) *Understanding Social Theory*. London: Sage.
Layder, D. (1997) Modern Social Theory: Key Debates and New Directions. London: UCL Press.
Layder, D. (1998a) Sociological Practice: Linking Theory and Social Research. London: Sage.
Layder, D. (1998b) 'The reality of social domains: Implications for theory and method', in T. May and M. Williams (eds), *Knowing the Social World*. Buckingham: Open University Press.
Layder, D. (2004) Emotion in Social Life: The Lost Heart of Society. London: Sage.
Layder, D. (2007) 'Self-identity and personhood in social analysis: The inadequacies of postmodernism and social constructionism', in J.L.Powell and T.Owen (eds), *Reconstructing Postmodernism: Critical Debates*. New York: Nova Science Publishers.

Layton, E.T. (1988) 'Science as a form of action. The role of the Engineering Sciences', *Technology and Culture*, 28: 594–607.
Lemert, C. (1993) 'After modernity', in C. Lemert (ed), *Social Theory: The Multicultural and Classical Readings*. San Francisco: Westview Press.
Lenoir, T. and Hays, M. (2000) 'The Manhattan project for biomedicine', in P.R. Sloan (ed), *Controlling Our Destinies: Historical, Philosophical, Ethical and Theological Perspectives on the Human Genome Project*. Notre Dame: Notre Dame Press.
Leuzinger, M. and Rambert, B. (1988) 'I can feel it-my baby is healthy: Women's experiences with *pre*-natal diagnosis in Switzerland', *Reproductive and Genetic Engineering*, 1(3): 239–49.
Levi, M. (2007) 'Organized crime and terrorism', in M. Maguire, R. Morgan and R. Reiner (eds), *The Oxford Handbook of Criminology* (4th Edition). Oxford: Oxford University Press.
Levi-Strauss, C. (1963) *Structural Anthropology*. New York: Basic Books.
Levi-Strauss, C. (1974) *The Savage Mind*. London: Weidenfeld and Nicholson.
Lewontin, R. (1992) 'The dream of the human genome', *New York Review of Books*, 39: 31–40.
Lewontin, R. (2000) *It Ain't Necessarily So*. New York: Granta Books.
Lippman, A. (1991) 'Prenatal genetic testing and screening: Constructing needs and reinforcing inequities', *American Journal of Law and Medicine*, 17: 15–50.
Loader, I. and Sparks, R. (2007) 'Contemporary landscapes of crime, order and control: Governance and risk and globalization', in M. Maguire, R. Morgan and R. Reiner (eds), *The Oxford Handbook of Criminology* (4th Edition). Oxford: Oxford University Press.
Lockwood, D. (1964) 'Social integration and system integration', in G.K. Zollschan and W. Hirsch (eds), *Explorations in Social Change*. London: Routledge.
Longino, C.F. and Powell, J.L. (2004) 'Embodiment and the study of ageing', in V. Berdayes (ed), *The Body in Human Inquiry: Interdisciplinary Explorations of Embodiment*. New York: Hampton Press.
Luhmann, N. (1982) *The Differentiation of Society*. New York: Columbia University Press.
Lupton, D. (1999) *Risk*. London: Routledge.
Lynn, R. (2001) *Eugenics: A Reassessment*. Westport, CT: Praeger.
Lyon, D. (1994) *Postmodernity*. Milton Keynes: Open University Press.
Lyotard, J.F. (1973) *La Fenomenologia*. Buenos Aires: EUDEBA.
Lyotard, J.F. (1984) *The Postmodern Condition: A Report on Knowledge*, trans. G. Bennington and B. Massumi. Manchester: Manchester University Press.
Lyotard, J.F. (1986–1987) 'Rules and paradoxes and svelte paradox', *Cultural Critique*, 5: 209–19.
MacInnes, J. (1998) *The End of Masculinity*. Buckingham: Open University Press.
MacKenzie, D. (1978) 'Statistical theory and social interests: A case study', *Social Studies of Science*, 8(1): 35–83.

MacKenzie, D. (1996) 'How do we know the properties of artefacts? Applying the sociology of knowledge to technology', in R. Fox (ed), *Technological Change: Methods and Themes in the History of Technology*. Reading: Harwood.

MacKenzie, D. and Wajcman, J. (eds) (1999) *The Social Shaping of Technology*. Buckingham: Open University Press.

Mannheim, K. (1925) 'The problem of a Sociology of knowledge from a dynamic standpoint', in H. Nowotny and K. Taschwer (eds), *The Sociology of the Sciences*, vol. 1: 3–14. heltenham: Edward Elgar.

Marsh, D. (ed) (1998) *Comparing Policy Networks*. Buckingham: Open University Press.

Marsh, D. and Smith, M. (2000) 'Understanding policy networks: Towards a dialectical approach', *Political Studies*, 48(1): 4–21.

Marsh, D. and Stoker, G. (eds) (1995) *Theory and Methods in Political Science*. London: Macmillan.

Martin, P. (1997) The Sickening Mind: Brain, Behaviour, Immunity and Disease. London: HarperCollins.

Martin, P. and Frost, R. (2003) 'Regulating the commercial development of genetic testing in the UK: Problems, possibilities and policy', *Critical Social Policy*, 23(2): 187–207.

McCormick, R.A. (1994) 'Blastomere separation: Some concerns', *Hastings Center Report*, 24(2) (March–April): 14–16.

McLaughlin, E. (2013) 'Serial killing', in E. McLaughlin and J. Muncie (eds), *The Sage Dictionary of Criminology* (3rd Edition). London: Sage.

McLennan, C. (1995) 'After postmodernism: Back to sociological theory?', *Sociology*, 29(1): 117–32.

McNay, L. (1992) Foucault and Feminism: Power, Gender and the Self. Cambridge: Polity Press.

May, T. (1996) *Situating Social Theory*. Milton Keynes: Open University Press.

Mead, G.H. (1967) *Mind, Self and Society*. Chicago: University of Chicago Press.

Mednick, S.A., Moffit, T. and Stack, S. (eds) (1987) *The Causes of Crime: New Biological Approaches*. Cambridge: Cambridge University Press.

Mednick, S.A. and Volavka, J. (1980) 'Biology and crime', in N. Morris and M. Tonry (eds), *Crime and Justice*, Vol.2. Chicago: University of Chicago Press.

Merriden, T. (2001) Irresistable Forces: The Legacy of Napster and the Growth of the Underground Internet. Oxford: Capstone.

Merton, R.K. (1938) *Science, Technology and Society in Seventeenth-Century England*. Bruges: St Catherine Press (4th Edition, with a new introduction. New York, Howard Fertig, 2001).

Merton, R.K. (1942) 'The normative structure of Science', *The Sociology of Science*, 1973.

Merton, R.K. (1952) 'Neglect of the Sociology of Science', in Storer, N.W. (ed) (1973) *The Sociology of Science*. Chicago: University of Chicago Press. pp. 210–23.

Merton, R.K. (1968) *Social Theory and Social Structure* (original edn. 1949). New York: Free Press.

Messerschmidt, J.W. (1993) *Masculinities and Crime*. Lantham, MD: Rowman and Littlefield.
Midgley, M. (2000) *Science and Poetry*. London: Routledge.
Mill, J.S. (1859) *On Liberty*. London: Macmillan.
Miller, S. (1994) 'Wrinkles, ripples and fireballs: Cosmology on the front page', *Public Understanding of Science*, 3(4): 445–53.
Milovanovic, D. (1996) 'Postmodern criminology', *Justice Quarterly*, 13(4): 567–609.
Milovanovic, D. (1997) *Chaos, Criminology and Social Justice*. New York: Praeger.
Milovanovic, D. (1999) 'Catastrophe theory, discourse and conflict resolution', in B. Arrigo (ed), *Social Justice/Criminal Justice*. Belmont, CA: Wadsworth.
Milovanovic, D. (2013) 'Postmodernism', in E. McLaughlin and J. Muncie (eds), *The Sage Dictionary of Criminology* (3rd Edition). London: Sage.
Mouzelis, N. (1989) 'Restructuring structuration theory', *The Sociological Review*, 37(4): 613–35.
Mouzelis, N. (1991) Back to Sociological Theory: The Construction of Social Order. London: Macmillan.
Mouzelis, N. (1993a) 'The poverty of sociological theory', *Sociology*, 27(4): 675–95.
Mouzelis, N. (1993b) 'Comparing the Durkheimian and Marxist traditions', *The Sociological Review*, 41(3): 572–82.
Mouzelis, N. (1994) 'In defence of "grand" historical sociology', *British Journal of Sociology*, 45(1): 31–6.
Mouzelis, N. (1995) Sociological Theory: What Went Wrong? Diagnosis and Remedies. London: Routledge.
Mouzelis, N. (1996) 'After postmodernism: A reply to Gregor McLennan', *Sociology*, 30(1): 131–5.
Mouzelis, N. (1997) 'Social and system integration: Lockwood, Habermas, Giddens', *Sociology*, 31(1): 111–19.
Mouzelis, N. (2000) 'The subjectivist-objectivist divide: Against transcendence', *Sociology*, 34(4): 741–62.
Mouzelis, N. (2007) 'Cognitive relativism: Between positivistic and relativistic thinking in the social sciences', in J.L. Powell and T. Owen (eds), *Reconstructing Postmodernism: Critica Debates*. New York: Nova Science Publishers.
Muncie, J. (2007) 'Globalization', in E. McLaughlin and J. Muncie (eds), *The Sage Dictionary of Criminology* (2nd Edition). London: Sage.
Munch, R. and Smelser, N. (1987) 'Relating the micro and macro', in J. Alexander, B. Giesen, R. Munch and N. Smelser (eds), *The Micro-Macro Link*. Berkeley, CA: University of California Press.
Nelken, D. (1997) 'The globalization of crime and criminal justice', *Current Legal Problems*, 50: 251–77.
Nelkin, D. (1994) 'Promotional metaphors: The gene in popular discourse', *Nature Review: Genetics*, 2: 555–9.

Nelkin, D. (2001) 'Anything for an edge: Breeding a race of champions by germline', *The Galton Institute Newsletter*, 43: 5–6.
Nesse, R. (1990) 'Evolutionary explanations of emotions', *Human Nature*, 1(3): 261–89.
Nesse, R.M. and Williams, G.C. (1994) *Evolution and Healing: The New Science of Darwinian Medicine*. London: Pheonix.
Newman, F. and Brody, E.B. (1988) 'Psychosocial and ethical concerns in new reproductive technologies', *Journal of Psychosomatic Obstetric Gynaecology*, 9: 155–8.
Newton, T. (2003) 'Truly embodied sociology: Marrying the social and the biological?', *Sociological Review*, 51(1): 20–41.
Nicholson, L. and Seidman, S. (1995) 'Introduction', in L. Nicholson and S. Seidman (eds), *Social Postmodernism: Beyond Identity Politics*. Cambridge: Cambridge University Press.
Nowotny, H. and Taschwer, K. (eds) (1996) *The Sociology of Sciences*, 2 vols. Cheltenham: Elgar.
O'Brien, M. (1981) *The Politics of Reproduction*. London: Routledge and Kegan Paul.
O'Brien, M. (1989) *Reproducing the World: Essays in Feminist Theory*. Boulder, CO: Westview Press.
Office of Population, Consensus and Surveys. (1993) 'Congenital malformation statistics. Her Majesty's Stationery Office', in R. Lynn (ed), *Eugenics: A Reassessment*. Westport, CT: Praeger.
Opitz, J.M. (2000) 'The geneticization of Western civilisation: Blessing or bane?', in P.R. Sloan (ed), *Controlling Our Destinies: Historical, Philosophical, Ethical, and Theological Perspectives on the Human Genome Project*. Notre Dame: Notre Dame Press.
Owen, T. (2006a) 'Towards a post-Foucauldian sociology of ageing', in J.L. Powell and A. Wahidin (eds), *Foucault and Ageing*. New York: Nova Science Publishers.
Owen, T. (2006b) 'Genetic-social science and the study of human biotechnology', *Current Sociology*, 54(6): 897–917.
Owen, T. (2007a) 'After postmodernism: Towards an evolutionary sociology', in J.L. Powell and T. Owen (eds), *Reconstructing Postmodernism: Critical Debates*. New York: Nova Science Publishers.
Owen, T. (2007b) 'Culture of crime control: Through a post-Foucauldian lens', *The Internet Journal of Criminology* (www.internetjournalofcriminology.com).
Owen, T. (2009a) *Social Theory and Human Biotechnology*. With a Foreword by Professor Derek Layder (University of Leicester). New York: Nova Science Publishers.
Owen, T. (2009b) 'England and Wales: The criminal justice system in "post-industrial society" ', in J.L. Powell and J. Hendricks (eds), *The Welfare State in Post-Industrial Society: A Global Perspective*. New York: Springer.
Owen, T. (2012a) 'The biological and the social in criminological theory', in S. Hall and S. Winlow (eds), *New Directions in Criminological Theory*. London: Routledge.

Owen, T. (2012b) 'Theorizing masculinities and crime: A genetic-social approach', *International Journal of Criminology and Sociological Theory*, 5(3): 972–84.

Owen, T. and Powell J.L. (2006) ' "Trust", professional power and social theory: Lessons from a post-Foucauldian framework', *International Journal of Sociology and Social Policy*, 26(3/4): 110–20.

Padgett, J.F. and Ansell, C.K. (1989) 'From faction to party in Renaissance Florence', Department of Political Science, University of Chicago, September 1989, p. 33 cited in White, H. (1992) *Identity and Control: A Structural Theory of Social Action*. Princeton, NJ: Princeton University Press, p. 292.

Parry, S. (2003) 'The politics of cloning: Mapping the rhetorical convergence of embryos and stem cells in parliamentary debates', *New Genetics and Society*, 22(2): 145–68.

Parsons, W. (1995) Public Policy: An Introduction to the Theory and Practice of Policy Analysis. Aldershot: Edward Elgar.

Paul, D. (1998) The Politics of Heredity: Essays on Eugenics, Biomedicine, and The Nature-Nurture Debate. Albany: State University of New York Press.

Peacocke, A.R. (2000) 'Relating Genetics to Theology on the map of scientific knowledge', in P.R. Sloan (ed), *Controlling Our Destinies: Historical, Philosophical, Ethical, and Theological Perspectives on the Human Genome Project*. Notre Dame: Notre Dame Press.

Pebley, A.R. and Westaff, C.F. (1982) 'Women's sex preferences in the United States: 1970 to 1975', *Demography*, 19: 177–90.

Pernick, M.S. (2000) 'Defining the defective: Eugenics, Esthetics, and Mass Culture in early twentieth century America', in P.R. Sloan (ed), *Controlling Our Destinies: Historical, Philosophical, Ethical, and Theological Perspectives on the Human Genome Project*. Notre Dame: Notre Dame Press.

Perrow, C. (1984) Normal Accidents: Living with High Risk Technologies. New York: Basic Books.

Peters, T. and Bennett, G.J. (2004) 'Science as saviour?' cloning, ethics and politics', *Dialog*, 43(3).

Petersen, A. and Bunton, R. (2002) *The New Genetics and the Public's Health*. London: Routledge.

Pickering, A. (1993) 'The mangle of practice: agency and emergence in the sociology of science', *American Journal of Sociology*, 99(3): 559–89.

Pickering, A. (1995a) 'Cyborg history and the World War II regime', *Perspectives on Science*, 3: 1–48.

Pickering, A. (1995b) *The Mangle of Practice: Time, Agency and Science*. Chicago: University of Chicago Press.

Pickering, A. (2001) *The Mangle of Practice: Time, Agency and Science* (2nd edition). Chicago: University of Chicago Press.

Pickering, A. and Cushing, J.T. (1986) 'Constructing quarks: a sociological history of particle physics', *American Journal of Physics*, 54: 581.

Pierre, J. and Peters, B.G. (2000) *Governance, Politics and the State*. London: Macmillan.

Pinker, S. (1994) The Language Instinct: The New Science of Language and Mind. London: Penguin.
Pinker, S. (1995) *The Language Instinct*. London: Penguin.
Pinker, S. (1997) *How the Mind Works*. New York: W.W. Norton.
Pinker, S. (1999) *How the Mind Works* (2nd Edition). London: Penguin.
Pinker, S. (2002) *The Blank Slate*. London: Penguin.
Pinker, S. and Rose, R. (1998) 'The Two Steves: Pinker vs Rose: A debate', *Edge*. Available at http//www.edge.org/3rd_culture/pinker_rose/pinker_rose_p1.html.
Plomin, R. (1997) *Behavioural Genetics*. London: W.H. Freeman.
Pokorski, R.J. (1997) 'Insurance underwriting in the genetic era', *American Journal of Human Genetics*, 60: 205–16.
Pollock, K. (1988) 'On the nature of social stress: Production of a modern mythology', *Social Science and Medicine*, 26(3): 381–92.
Powell, J. (1988) '"The Us and the Them": Connecting Foucauldian and political economy insights into aging bodies'. Paper presented to the British Sociological Association Annual Conference, University of Edinburgh.
Powell, J.L. (2001) 'Social theory and the ageing body', *International Journal of Language, Society and Culture*, 8(2): 2–16.
Powell, J.L. (2006) *Rethinking Social Theory and Later Life*. New York: Nova Science Publishers.
Powell, J.L. and Biggs, S. (2000) 'Managing old age: The disciplinary web of power, surveillance and normalisation', *Journal of Aging and Identity*, 5(1): 3–13.
Powell, J.L. and Biggs, S. (2001) 'Genealogies of old age and social welfare', *Journal of Aging and Social Policy*, 12(2): 93–115.
Powell, J.L. and Biggs, S. (2004) 'Ageing, technologies of self and bio-medicine: a Foucauldian Excursion', *International Journal of Sociology and Social Policy*, 25(13): 96–115.
Powell, J.L. and Owen, T. (2005) 'The bio-medical model and aging: Towards an anti-reductionist model?', *International Journal of Sociology and Social Policy*, 25(9): 27–40.
Quilley, S. and Loyal, S. (2005) 'Eliasian sociology as a "central theory" for the human sciences', *Current Sociology*, 53(5): 807–28.
Radcliffe Richards, J. (2001) Human Nature after Darwin: A Philosophical Introduction. London: Routledge.
Ratzinger, J. and Bovone, A. (1987) 'Instruction on respect for human life in its origin and on the dignity of procreation'. Congregation for the Doctrine of the Faith, the Feast of the Chair of St. Peter, the Apostle, Rome, February 22nd (1987), cited in R.Lynn etc.
Rechsteiner, M. (1991) 'The human genome project: Misguided science policy?', *Trends in Biochemical Sciences*, 16: 453–59.
Reiner, R. (2012) 'Political economy and criminology: The return of the repressed', in S. Hall and S. Winlow (eds), *New Directions in Criminological Theory*. London: Routledge.

Renn, O. (1992) 'Concepts of risk: A classification', in S. Krimsky and D. Golding (eds), *Social Theories of Risk*. Westport, CT: Praeger.

Revelli, A., Tur-Kaspa, I., Holte, G.J. and Massobrio, M. (eds) (2003) *Biotechnology of Human Reproduction*. London: Taylor and Francis.

Rhodes, R.A.W. (1997) Understanding Governance: Policy Networks, Governance, Reflexivity and Accountability. Buckingham: Open University Press.

Rich, A. (1981) 'Disobedience is what NWSA is potentially about', *Women's Studies Quarterly*, 9(3): 4–5.

Richardson, J. (1996) 'Policy-making in the EU: Interests, ideas, and garbage cans of primeval soup', in J. Richardson (ed), *European Union: Power and Policy Making*. London: Routledge.

Ridley, M. (1999) Genome: The Autobiography of a Species in 23 Chapters. London: Fourth Estate.

Ridley, M. (2003) Nature Via Nurture: Genes, Experience and What Makes Us Human. London: Fourth Estate.

Ritzer, G. (1990) 'Micro-macro linkages in sociology; applying a metatheoretical tool', in G. Ritzer (ed), *Frontiers of Social Theory; The New Syntheses*. New York: Columbia University Press.

Ritzer, G. (1992) 'Metatheorising in sociology: Explaining the coming of age', in G. Ritzer (ed), *Metatheorizing*. London: Sage.

Ritzer, G. (2000) *Modern Sociological Theory*. New York: McGraw-Hill.

Ritzer, G. and Ryan, J.M. (2007) 'Postmodern social theory and sociology: On symbolic exchange with a "dead" theory', in J.L. Powell and T. Owen (eds), *Reconstructing Postmodernism: Critical Debates*. New York: Nova Science Publishers.

Robbe-Grillet, A. (1963) *For a New Novel: Essays on Fiction*. Paris: Les Editions de Minuit.

Roberts, D.E. (1996) 'Race and the new reproduction', *Hastings Law Journal*, 47: 935–49.

Rojek, C. and Turner, B. (2000) 'Decorative sociology: Towards a critique of the cultural turn', *The Sociological Review*, 48(4): 629–48.

Rogers, L. (4 July 1999) 'Disabled children will be a "sin", says scientist', *Sunday Times* (London), p. 15, in R. Lynn (2001) *Eugenics: A Reassessment*. Westport, Connecticut: Praeger.

Rose, H. (2000) 'Colonising the social sciences?', in H. Rose and S. Rose (eds), Alas Poor Darwin: Arguments Against Evolutionary Psychology. London: Jonathan Cape.

Rose, H. and Rose, S. (eds) (2000) Alas, Poor Darwin: Arguments Against Evolutionary Psychology. London: Jonathan Cape.

Rose, N. and Miller, P. (1992) 'Political power beyond the state: Problematics of government', *British Journal of Sociology*, 43(2): 173–205.

Rose, S. (ed) (1982) *Towards a Liberatory Biology*. London, New York: Allison and Busby.

Rose, S. (1997) Lifelines: Biology, Freedom, Determinism. London: Penguin.

Rose, S. (ed) (1999) From Brains to Consciousness: Essays on the New Science of the Mind. London: Penguin.

Rose, S. (2000) 'Escaping evolutionary psychology', in H. Rose and S. Rose (eds), *Alas, Poor Darwin: Arguments Against Evolutionary Psychology*. London: Jonathan Cape.

Rose, S., Kamin, L.J. and Lewontin, R.C. (1984) *Not in Our Genes*. London: Pantheon.

Royal College of Physicians. (1989) *Prenatal Diagnosis and Genetic Screening*. London: Royal College of Physicians, in R. Lynn (2001) *Eugenics: A Reassessment*. Westport, CT: Praeger.

Rubinsztein, D., Leggo, J. and Coles, R. (1996) 'Phenotypic characterization of individuals with 30–40 CAG repeats in the Huntington Disease (HD) gene reveals HD cases with 36 repeats and apparently normal elderly individuals with 36–39 repeats', *American Journal of Human Genetics*, 59: 16–22, in P.R. Sloan (eds) (2000) *Controlling Our Destinies: Historical, Philosophical, Ethical, and Theological Perspectives on the Human Genome Project*. Notre Dame: Notre Dame Press.

Rubenstein, D.S., Thomasma, D.C., Schon, E.A. and Zinaman, M.J. (1995) 'Germ-line therapy to cure mitochondrial disease: Protocol and ethics of in vitro ovum nuclear transplantation', *Cambridge Quarterly of Healthcare Ethics*, 4(3): 316–39.

Russell, A. and Vogler, J. (2001) *The International Politics of Biotechnology: Investigating Global Futures*. Manchester: Manchester University Press.

Russett, B. and Starr, H. (1996) *World Politics: The Menu for Choice*. New York: Freeman and Company.

Sahlins, M.D. (1972) *Stone Age Economics*. Chicago: Aldine-Atherton.

Saussure, F. de ((1916)1974) *Course in General Linguistics*, trans. W. Baskin. London: Collins.

Scambler, G. (2000) *Health and Social Change: A Critical Theory*. Buckingham: Open University Press.

Schaffner, K.F. (2000) 'Reductionism and determinism in human genetics: Lessons from simple organisms', in P.R. Sloan (ed), *Controlling Our Destinies: Historical, Philosophical, Ethical, and Theological Perspectives on the Human Genome Project*. Notre Dame: Notre Dame Press.

Schleiermacher, S. (1990) 'Racial hygiene and "Deliberate Parenthood": Two sides of demographer Hans Harmsen's population policy', *Issues in Reproductive and Genetic Engineering: Journal of International Feminist Analysis*, 3(3): 201–10.

Scholes, R. (1974) *Structuralism in Literature*. New Haven, CT: Yale University Press.

Scruton, R. (1994) *Modern Philosophy: An Introduction and Survey*. London: Arrow.

Seidman, S. (1992) 'Postmodern social theory as narrative with a moral intent', in S. Seidman and D. Wagner (eds) *Postmodernism and Social Theory*. Oxford: Blackwell.

Seidman, S. (1994) *Contested Knowledge: Social Theory in the Postmodern Era*. Oxford: Blackwell.

Shapin, S. (1982) 'History of science and its sociological reconstructions', *History of Science*, 20: 157–211.
Sharp, R. (1980) Knowledge, Ideology and the Politics of Schooling: Towards a Marxist Analysis of Schooling. London: Routledge.
Shaw, C.R. and McKay, H.D. (1942) *Juvenile Delinquency and Urban Areas*. Chicago: University of Chicago Press.
Shaw, M.W. (1984) 'To be or not to be? That is the question', *American Journal of Human Genetics*, 36, 1–9.
Shilling, C. (1993) *The Body and Social Theory*. London: Sage.
Shiva, V. (1988) Staying Alive: Women, Ecology and Development. London: Zed.
Sibeon, R. (1996) Contemporary Sociology and Policy Analysis: The New Sociology of Public Policy. London: Kogan Page and Tudor.
Sibeon, R. (1997a) 'Power agency/structure and micro-macro; an excursus in anti-reductionist sociology', Paper presented at Annual Conference of the British Sociological Association, University of York (April).
Sibeon, R. (1997b) 'Anti-reductionist sociology and the study of postnational governance', Paper presented at the Third Conference of the European Sociological Association, University of Essex (September).
Sibeon, R. (1999) 'Anti-reductionist sociology', *Sociology*, 32(2): 317–34.
Sibeon, R. (2001) 'The Reconstruction of sociological theory', Paper presented at the 50th Anniversary Conference of the British Sociological Association, Manchester, 9–12 (April).
Sibeon, R. (2004) *Rethinking Social Theory*. London: Sage.
Sibeon, R. (2007) 'An excursus in post-postmodern social science', in J.L. Powell and T. Owen (eds), *Reconstructing Postmodernism: Critical Debates*. New York: Nova Science Publishers.
Singer, E. (1991) 'Public attitudes toward genetic testing', *Population Research and Policy Review*, 10: 235–55.
Sloane, P.R. (ed) (2000) Controlling Our Destinies: Historical, Philosophical, Ethical, and Theological Perspectives on the Human Genome Project. Notre Dame: Notre Dame Press.
Smart, B. (2007) '(Dis) interring postmodernism or a critique of the political economy of consumer choice', in J.L. Powell and T. Owen (eds), *Reconstructing Postmodernism: Critical Debates*. New York: Nova Science Publishers.
Smart, C. (1995) *Law, Crime and Sexuality*. London: Sage.
Soja, E. (1989) Postmodern Geographies: The Reassertion of Space in Critical Theory. London: Verso.
Sokal, A. (1996) 'Transgressing the boundaries: Towards a transformative hermeneutics of quantum gravity', *Social Text*, 14(1–2): 217–52.
Sokal, A. and Bricmont, J. (1987) 'Impostures intellectuelles, Paris, Editions Odile Jacob' (English trans. Sokal, A. and Bricmont, J. (1998)), *Fashionable Nonsense: Postmodern Intellectuals' Abuse of Science*. New York: Picador.
Stones, R. (1996) Sociological Reasoning: Towards a Past-Modern Sociology. London: Macmillan.

Sturgis, P., Cooper, H. and Fife-Schaw, C. (2005) 'Attitudes to Biotechnology: Estimating the opinions of a better-informed public', *New Genetics and Society*, 24(1): 31–56.

Sulston, J. and Ferry, G. (2003) *The Common Thread: Science, Politics, Ethics and the Human Genome*. London: Corgi.

Sutherland, E.H. (1947) *Principles of Criminology*. Philadelphia, PA: J.B. Lippincott.

Suzuki, D. and Knudtson, P. (1990) 'Genetics: The clash between the new genetics and human values', Cambridge, MA: Harvard University Press, in R. Lynn (2001) *Eugenics: A Reassessment*. Westport, CT: Praeger.

Swingewood, A. (2000) *A Short History of Sociological Thought*. London: Macmillan.

Symons, D. (1979) *The Evolution of Human Sexuality*. New York: Oxford University Press.

Symons, D. (1995) 'Beauty is in the adaptations of the beholder: The evolutionary psychology of human female sexual attractiveness', in P.R. Abramson and S.D. Pinkerton (eds), *Sexual Nature, Sexual Culture*. Chicago: University of Chicago Press.

Tattersall, I. (1998) *Becoming Human*. New York: Harcourt Brace.

Tattersall, I. (2003) 'Email correspondence', in M. Ridley (ed), *Nature via Nurture: Genes, Experience and What Makes Us Human*. London: Fourth Estate.

Tawber, A.I. and Sarker, S. (1992) 'The human genome project: Has blind reductionism gone too far?', *Perspectives in Biology and Medicine*, 35: 220–35.

Taylor, S. (2006) The Sociology of Emotion: An Introduction. London: Palgrave.

Testart, J. (1995) 'The new eugenics and medicalised reproduction', *Cambridge Quarterly of Health Care Ethics*, 4, 304–12.

Timson, J. (1997) 'Human genetics', in R.A. Peel (ed), *Essays in the History of Eugenics*. Proceedings of a conference organised in London: The Galton Institute.

Thain, M. and Hickman, M. (1995) *The Penguin Dictionary of Biology*. London: Penguin.

Tomsen, S. (ed) (2008) *Crime, Criminal Justice and Masculinities*. Sydney: University of Western Sydney.

Tooby, J. (1999) 'The most testable concept in Biology, Part 1', *HBES Newsletter* (Fall). Available at http://www.psych.ucsb.edu/research/cep/viewfall99.html.

Tooby, J. and Cosmides, L. (1990) 'The past explains the present: Emotional adaptations and the structure of ancestral enviornments', *Ethnology and Sociobiology*, 11: 375–424.

Tooby, J. and Cosmides, L. (1992) 'The psychological foundations of culture', in J.H. Barkow, L. Cosmides and J. Tooby (eds), *The Adapted Mind: Evolutionary Psychology and the Generation of Culture*. New York: Oxford University Press.

Tooby, J. and De Vore, I. (eds) (1987) 'The reconstruction of hominid behavioural evolution through strategic modelling', in *The Evolution of Human Behaviour: Primate Models*. Albany, NY: SUNY Press.
Touraine, A. (1981) *The Voice and the Eye: An Analysis of Social Movements*. Cambridge: Cambridge University Press.
Touraine, A. (1995) *Critique of Modernity*. Oxford: Blackwell.
Trivers, R. (1981) 'Sociobiology and politics', in E. White (ed), *Sociobiology and Human Politics*. Lexington, MA: Heath and Company.
Turner, B.S. (1990) 'Conclusion: Peroration on ideology', in N. Abercrombie, S. Hill and B.S. Turner (eds), *Dominant Ideologies*. London: Unwin Hyman.
Turner, B.S. and Rojek, C. (2001) Society and Culture: Principles of Scarcity and Solidarity. London: Sage.
Turner, L. (2004) 'Biotechnology as religion', *Nature Biotechnology*, 22(6): 604–59.
Van Gelder, T. (1995) 'What might cognition be, if not computational?', *Journal of Philosophy*, 91(7): 345–81.
Van Loon, J. (2002) Risk and Technological Culture: Towards a Sociology of Violence. London: Routledge.
Verlinsky, Y., Pergament, E. and Strom, C. (1990) 'The preimplantation diagnosis of genetic diseases', *Journal of In Vitro Fertilization and Embryo Transfer*, 7, 1–5.
Wade, N. (1998) 'It's a three-legged race to decipher the human genome', *New York Times*, (23 June): 811.
Wahidin, A. (2003) 'Reclaiming agency: Managing ageing bodies in prison', in E. Tulle (ed), *Old Age and Human Agency*. New York: Nova Science Publishers.
Wainwright, D. and Calman, M. (2002) *Work Stress: The Making of a Modern Epidemic*. Buckingham: Open University Press.
Wakayama, T., Perry, A.C., Zuccotti, M., Johnson, K.R. and Yanagimachi, R. (1998) 'Full-term development of mice from enucleated oocytes injected with cumulus cell nuclei', *Nature*, 394: 369–74.
Walby, S. (1990) *Theorising Patriarchy*. Oxford: Blackwell.
Walker, A. and Shipman, P. (1996) *The Wisdom of the Bones: In Search of the Human Origins*. London: Weidenfeld and Nicolson.
Walklate, S. (2007) *Understanding Criminology: Current Theoretical Debates* (3rd Edition). London: McGraw Hill.
Walsh, A. and Beaver, K.M. (eds) (2009) *Biosocial Criminology: New Directions in Theory and Research*. New York: Routledge.
Walsh, A. and Ellis, L. (eds) (2003) *Biosocial Criminology: Challenging Environmentalism's Supremacy*. New York: Nova Science Publishers.
Warnock, M. (1987) 'Do human cells have rights?', *Bioethics*, 1: 8.
Weatherall, D.J. (1991) *The New Genetics and Clinical Practice*. Oxford: Oxford University Press, in R. Lynn (2001) *Eugenics: A Reassessment*. Westport, CT: Praeger.
Webster, A. (2005) 'Social science and a post-genomic future: alternative readings of genomic agency', *New Genetics and Society*, 24(2): 227–38.

Wertz, D.C. (1998) 'Eugenics is alive and well: A survey of genetic professionals around the world', *Science in Context*, 11: 493–510.
West, M. (2003) *The Immortal Cell: One Scientist's Quest to Solve the Mystery of Human Aging*. New York: Doubleday.
White, H.C. (1992) *Identity and Control: A Structural Theory of Social Action*. Princeton, NJ: Princeton University Press.
Wieviorka, M. (2012) 'From social order to the personal subject: A major reversal', in S. Hall and S. Winlow (eds), *New Directions in Criminological Theory*. London: Routledge.
Wilkie, T. (1994) *Perilous Knowledge: The Human Genome Project and Its Implications*. London: Faber and Faber Ltd.
Wilkinson, R.G. (1996) *Unhealthy Societies: The Affliction of Inequality*. London: Routledge.
Wilkinson, R.G. (2000a) *Mind the Gap: Hierarchies, Health and Human Evolution*. London: Weidenfeld and Nicolson.
Wilkinson, R.G. (2000b) 'Deeper than "neoliberalism": A reply to David Coburn', *Social Science and Medicine*, 51: 997–1000.
Williams, S.J. (1996) 'The vicissitudes of embodiment across the chronic illness trajectory', *Body and Society*, 2(2): 23–47.
Williams, S.J. (1998) ' "Capitalising" on emotions? Rethinking the inequalities in health debate', *Sociology*, 32(1): 121–39.
Williams, S.J. (2003) 'Marrying the social and the biological? A rejoinder to Newton', *Sociological Review*, 51(4): 550–61.
Williams, S.J. and Bendelow, G. (1998) *The Lived Body: Sociological Themes Embodied Issues*. London: Routledge.
Williams-Jones, B. and Graham, J.E. (2003) 'Actor-network theory: A tool to support ethical analysis of commercial genetic testing', *New Genetics and Society*, 22(3): 271–96.
Wilmut, I., Schnieke, A., McWhis, J., Kind, A. and Campbell, K.H. (1997) 'Viable offspring derived from fetal and adult mammation cells', *Nature*, 390: 27.
Wilson, D. (2012) 'Late capitalism, vulnerable populations and violent predatory crime', in S. Hall and S. Winlow (eds), *New Directions in Criminological Theory*. London: Routledge.
Wilson, J.Q. and Herrnstein, R.J. (1985) *Crime and Human Nature*. New York: Simon and Schuster.
Winlow, S. (2001) *Badfellas: Crime, Tradition and New Masculinities*. Oxford: Berg.
Winlow, S. and Hall, S. (2009) 'Retaliate first: Memory, humiliation and male violence', *Crime, Media, Culture*, 5(2): 285–304.
World Health Organisation. (11 March 1997). Press release on cloning, 20, in R. Lynn (2001) *Eugenics: A Reassessment*. Westport, CT: Praeger.
Yar, M. (2012) 'Critical criminology, critical theory and social harm', in S. Hall and S. Winlow (eds), *New Directions in Criminological Theory*. London: Routledge.

Yeates, N. (2001) *Globalization and Social Policy.* London: Sage.
Young, A. (1995) *Imagining Crime.* London: Sage.
Young, J. (2013) 'Left realism', in E. McLaughlin and J. Muncie (eds), *The Sage Dictionary of Criminology* (3rd Edition). London: Sage.
Ziman, J. (2001) *Real Science. What It Is, and What It Means.* Cambridge: Cambridge University Press.
Zimmerman, S. (1990) 'Industrial capitalism's "hostility to childbirth", "responsible childbearing" and eugenic reproductive politics in the first third of the 20th century', *Issues in Reproductive and Genetic Engineering: Journal of International Feminist Analysis,* 3(3): 191–200.

Index

abolitionism, 17, 159
actor-network theory
  biological sciences, 122–6
  generally, 16, 32, 44
  reification, 123–30
  rejection, 122–6
agency
  agency-structure distinction, 44
  human, 34, 128
  non-human, 127, 170
  post-humanism, 126–30
  reification, 15–17, 38
  social actors, 33–45
  see also non-agency
agency-structure, 30–40
Alas, Poor Darwin: Arguments against Evolutionary Psychology (Rose, H., Rose, S.), 78, 81, 93
Albrow, M., 15
Alexander, J., 29
Anderson, B., 139, 140
Anderson, J.L., 87
Ansell, C.K., 131
anti-reductionism, 5, 11, 35
Archer, M., 1, 6, 8, 10, 13, 18, 29, 33, 34, 35, 40, 41, 42, 43, 48, 49, 50, 51, 52, 54, 59, 60, 61, 96, 98, 101, 114, 115, 121, 164, 165, 166
Armstrong, D., 31, 68
authoritarian populism, 8, 18, 134

Barkow, J., 74, 76, 77, 146
Barrett, L., 75
Barthes, R., 37
Baudrillard, J., 21
Bauman, Z., 22
Beaver, K.M., 1, 4, 7, 65, 158
Becker, J.B., 3

behavioural genetics, human behaviour, 7, 12, 63, 102, 121, 153
Bell, D., 140
Bendelow, G., 105
Benjamin, J., 145
Bennington, J., 161
Benton, T., 12, 66, 67, 68, 69, 70, 71, 73, 76, 95, 96, 97, 103, 104, 145, 154, 165
Berger, P., 24
Best, S., 23, 32
Betts, K., 11, 17, 65, 117, 133
Bhaskar, R., 50
Biggs, S., 68
biological variable
  basis for human behaviour, 121
  defined, 12
  element in criminal behaviour, 3–4
  generally, 153–9
  metatheoretical framework incorporation, 2, 8, 101–15
biophobia, 7, 66, 173
biosocial criminology, 7
Birke, L., 73
Blackledge, D., 19
Bogaert, A.F., 107, 114, 122, 142, 153, 157
Bourdieu, P., 39, 53
Bretherton, C., 139
Brittan, S., 110
Brown, S., 128, 170
Brunner, H.G., 144
Bucchi, M., 124
building bridges, 65–73
Bulmer, M., 120
Bunton, R., 76
Bury, M.R., 12, 69, 154
Buss, D.M., 76, 82, 89

Calhoun, C., 26
Callon, M., 16, 32, 42, 124, 125, 126, 129, 160, 161
Cannon, W.B., 103, 104, 114, 145
Canter, D., 168, 169
cardinal sins
  defined, 5–6, 12–18, 63
  essentialism, 14–15, 131–3
  functional teleology, 17–18, 133–5
  reductionism, 13–14, 135–6
  reification, 15–17, 123–30
catecholamine, 72
Cavadino, M., 140
central conflation, *see* conflation
Chicago Theory, 153
Churchland, S., 90
Cicourel, A.V., 41
Clarke, J., 134
Coburn, D., 72
Cohen, I., 20, 27, 36
Coleman, R., 160
Collier, R., 155, 157, 158
Collins, H.M., 79, 80
conflation
  central, 49–51, 98–100, 121, 165
  types, 49–51, 54, 60
Connell, R., 18, 155, 156, 158
Copeland, P., 2, 8, 31, 101, 106, 107, 108, 110, 111, 112, 113, 114, 118, 122, 141, 145, 148, 151, 153, 156, 157
cortisol
  catecholamine, 72
  criminal behaviour, 158
  gene switching, 3, 149–50
  immune system, 3
Cosmides, L., 2, 82, 84, 88, 89, 90, 114, 142, 146, 147, 155, 170
Crawford, C.B., 87
criminal behaviour, 158
criminological and social theory, transitions in, 10–62
critical criminology, 4, 17, 159
cultural Turn
  generally, 5–6, 25, 30
  nihilism, 1, 33, 158
  reaction, 10, 66–7
  see also relativism
culture of crime control, 32
Curry, O., 7, 12, 76, 78, 81, 82, 83, 84, 85, 86, 87, 88, 89, 90, 91, 92, 93, 94, 114, 142, 155
cyber-crime, 8, 128, 170

Daly, M., 7, 12, 76, 84, 91, 92, 94, 155
Danaher, G., 16
Darwin, C., 90
David, M., 7, 76, 79, 80, 155
Davies, I., 25
Dawkins, R., 82, 105, 155
De Vore, I., 7, 12, 84, 155
Delanty, G., 16, 122
Derrida, J., 37
Dickens, P., 71, 76, 103
Dignan, J., 140
Dobash, R.E., 131
Dobash, R.P., 131
Dominelli, L., 14
dualism
  analytical, 30, 51
  generally, 98–105, 164–6
duality *vs.* dualism
  generally, 42, 59–60, 120
  revisited, 95–101
duality of structure, 42–5, 96–101, 164–5
Dunbar, R.I.M, 75, 76, 113, 114, 122, 142, 148, 153, 157
Dupre, J., 76
Durkheim, E., 17, 22, 34, 59, 60, 120
Dye, T., 47

Elias, N., 35
Elliot, P., 132
Ellis, L., 1, 4, 7, 65, 158
Elstad, J.I., 72
Enard, W., 8, 113, 114, 141, 153, 157
EP, *see* evolutionary psychology

essentialism
  defined, 14
  see also cardinal sins
Ettorre, E., 76
evolutionary psychology
  behavioural genetics, 63, 121, 141-2
  see also nature via nurture
The Expression of the Emotions in Man and Animals (Darwin, C.), 90

Farganis, S., 136
Feminism and Feminist Criminologies
  essentialism, 131-4
  generally, 20-1, 24, 93
Ferrell, J., 10
Filley, C.M., 3
Fisher, W.A., 107, 114, 122, 142, 153, 157
Fodor, J., 7, 77, 78, 79, 81
Foucauldian, 2-8, 30-4, 47-8, 54-5, 63-4, 68, 102, 106, 114, 131, 135, 156, 161-5, 172. see also post-Foucauldian
  power, 31-3, 119-20, 159-60
Foucault, M., 15-18, 30-3, 37, 42, 44, 48, 63, 106, 119, 121-2, 130-1, 135, 151-4, 161-4, 172
Fox, C.J., 22
Fox, N.J., 22
Freese, J., 1, 7, 66
Freund, P.E.S., 69, 96, 104, 115
Friedrichs, R.W., 27
functional teleology
  defined, 17
  see also cardinal sins

Gagnon, J.H., 106, 108, 114, 118, 122, 151, 152, 154
Garland, D., 101, 130, 160, 162, 164, 167, 172
Gellner, E., 10, 22
gene switching, 3, 149-50

genetic fatalism, 1-8, 11-12, 23, 63-5, 68, 70, 75, 88, 95, 102-5, 108, 114, 123, 136, 153, 154, 158
  meta-concept defined, 141-7
genetic-social
  agency-structure, micro-macro and time-space, 169-72
  application to study of crime and, 116-73
  biological variable, 153-9
  constructing, 63-115
  dualism, 164-6
  essentialism, 131-3
  functional teleology, 133-5
  genetic fatalism, 141-7
  metatheoretical framework, 116-22
  oversocialised gaze, 147-53
  power, 159-64
  psychobiography, 166-9
  reductionism, 135-41
  reification, 123-30
genetic-social science, 64
Genome: The Autobiography of a Species in 23 Chapters (Ridley), 108
Gibbs, J.P., 26
Giddens, A., 27, 34, 40, 42, 43, 46, 50, 74, 97, 98, 102, 108, 114, 147
Gigerenzer, G., 77
globalization, 135-41
Goffman, E., 56, 57
Grabosky, P.N., 128
Graham, J.E., 124, 125, 126
Grush, R., 90
Gurnah, A., 20, 26, 28

Habermas, J., 38, 39, 48, 59
Hall, S.S., 10, 18, 134, 155, 158
Hamer, D., 2, 8, 31, 101, 106, 107, 108, 110, 111, 112, 113, 114, 118, 122, 141, 145, 148, 151, 153, 156, 157
Hanmer, J., 132, 135

Harre, R., 15, 117, 169, 170
Harris, J.R., 108, 109, 114, 122, 142
Harvey, D., 21
Harvey, J., 161
Hay, C., 22
Herrnstein, R.J., 4, 158
Hester, M., 131
Hickman, M., 86
Hindess, B., 11, 13, 15, 34, 35, 64, 117, 119, 135, 161, 169, 170
Hochschild, A.R., 66, 67, 68
Holmwood, J., 10, 11
Holzner, B., 171
Homicide (Wilson, Daly), 92
Hopkins Burke, R., 134, 144
human biotechnology
 advances, 154
 gene patenting, commercial genetic testing, 124
 generally, 157, 167
 study of, 7, 12, 116
Human Genome Project, 75, 105, 143, 146–8, 154, 161, 170
Human Genome Sequencing Corporation, 46
Hunt, B., 19

immune system, 3
Irwin, A., 76

James, S., 47
Jameson, F., 21
Jefferson, T., 155, 156, 157, 158
Jessel, T., 143
Jones, M., 17, 126, 127, 128, 129

Kamin, L.J., 142, 155
Kandel, E., 143
Karmiloff-Smith, A., 78
Katz, S., 68
Kellner, D., 23, 32
Kelsoe, J., 144
Kennedy, J., 144
Klinteberg, B., 3, 151
Knorr-Cetina, K.D., 35, 41
Kooiman, J., 47

Kumar, K., 21
Kupferman, I., 143
Kusch, M., 79, 80

Lacan, J., 37
Laclau, E., 38
Lai, C.S.L., 113, 114, 141–2
Latour, B., 16, 32, 42, 124, 125, 126, 128, 161
Law, J., 24, 44, 124, 160, 161
Layder, D., 1, 2, 6, 10, 18, 33, 34, 43, 44, 48, 49, 54, 55, 56, 57, 58, 59, 61, 65, 66, 101, 120, 121, 123, 159, 163, 164, 165, 166, 167
Left Idealism, 123, 130, 159–60
Left Realism, 130, 159
Lemert, C., 11, 22
Levi, M., 138
Levi-Strauss, C., 36
Lewontin, R.C., 70, 75, 141, 142, 155
Loader, I., 139
Lockwood, D., 45
Loyal, S., 7, 67
Luckmann, T., 24
Lyon, D., 21, 22
Lyotard, J.F., 5, 11, 19, 20, 21, 24, 25, 34

MacInnes, J., 155, 157, 158
Mandell, N., 132
Marsh, D., 13, 22
Martin, P., 3, 94, 158
Marxist criminology, 8
masculinities and crime, 155–8, 166–7
McKay, H.D., 153
McLaughlin, E., 167, 168, 169
McLennan, C., 10, 13, 39, 40
Mead, G.H., 53
Mednick, S.A., 4, 158
Messerschmidt, J.W., 155, 156, 157, 158
meta-concepts, *see* agency-structure

meta-concepts, new, *see* biological variable; genetic fatalism; oversocialised gaze
metatheoretical
metatheoretical framework, codification of, 64–5
meta-theory, 119–20
metatheory, problems of relativism, 18
micro-macro, 30–5, 39–49, 60–5, 164–9
Midgley, M., 75, 76
Miller, H.T., 22
Milovanovic, D., 5, 20–2
The Mind Doesn't Work That Way: The Scope and Limits of Computational Psychology (Fodor), 79
Mouffe, C., 38
Mouzelis, N., 1, 6, 10, 13, 18, 19, 25, 28, 31, 33, 34, 39, 40, 43, 45, 46, 47, 48, 49, 52, 53, 54, 59, 60, 61, 64, 98, 99, 100, 153, 164, 166
Munch, R., 35, 41
Muncie, J., 137, 138, 140
Murray, C., 4, 117, 158

nature via nurture
  evolutionary psychology, 73–95
  generally, 7, 145, 165
Nelken, D., 137
neo-functionalism, 10
Nesse, R.M., 85
Newton, T., 12, 42, 69, 70, 72, 95, 96, 97, 102, 103, 104, 105, 115, 145, 148, 149, 150, 154, 165
Nicholson, L., 20
nihilistic relativism, 1, 158
non-agency, 123
norepinephrine, 2–3, 148–51

ontological flexibility, 27, 58
Opitz, J.M., 154
oversocialised gaze, 1–8, 11–12, 105–6, 151–4
  defined, 147

Owen, T., 1, 2, 4–7, 7–75, 10, 12, 19, 23, 30–4, 48–9, 52, 54, 63–8, 70, 84–5, 88, 94, 101, 106, 115–16, 118, 121–3, 130, 136–8, 141, 145, 147, 150–3, 155, 157–60, 162, 165, 166–7, 170, 172–3

Padgett, J.F., 131
Parsons, W., 47, 49, 60
Peters, B.G., 47, 76
Petersen, A., 76
Pickering, A., 16, 126, 127, 129, 170
Pierre, J., 47
Pinker, S., 2, 7, 73, 74, 76, 77, 79, 89, 90, 111, 114, 142, 148, 153, 157
Policing the Crisis: Mugging, the State and Law and Order (Hall, Critcher, Jefferson), 134, 158
Pollock, K., 95, 96, 148, 149, 150, 153, 157
post-Foucauldian, 12, 64, 121
post-humanism, *see* agency
postmodern criminology, 21–2
postmodernism, 4, 13–14, 20–1
post-postmodern
  generally, 1, 5–6
  social theory, 13, 18, 22
  see also genetic-social
post-postmodern theorists, 49–61
post-structuralism, 44, 48
Powell, J.L., 12, 23, 64, 65, 67, 68, 88, 121, 122, 141, 152, 153, 155, 159, 160
power, 31–2, 120, 159, 163–4, 168
Principles of Neural Science (Kandel, Schwartz, Jessell), 143
psychobiography, 55–6, 61, 65–6, 101, 156, 163–9

Quilley, S., 7, 67

reductionism
  defined, 13
  see also cardinal sins

reification, defined 15
  actor-network theory, 123–30
  agency, 15–17, 38
  genetic-social, 123–30
  see also cardinal sins
Reiner, R., 10
relativism
  anti-foundational, 4–5, 8, 64
  Foucauldian analysis, 160
  post-modernism, 21
Rhodes, R.A.W., 13, 22, 47
Richardson, J., 47
Ridley, M., 7, 8, 65, 67–8, 70–1, 73–5, 79–82, 88, 93–7, 101, 102, 104–5, 108–9, 110, 113–14, 118, 121–3, 136, 141–2, 145–51, 153–5, 157, 165
Ritzer, G., 13, 20, 35, 41
Robbe-Grillet, A., 4, 135
Rojek, C., 13, 24, 25
Rose, H., 77–8, 81, 85, 91–3
Rose, S., 78, 83, 85–6, 88–91, 93, 142, 155
Rubinsztein, D., 144
Russett, B., 46
Ryan, J.M., 20

Sahlins, M.D., 95, 96, 148
Saussure, F. de, 36
Scambler, G., 72–3
Schaffner, K.F., 142, 143, 144, 145, 154
Scholes, R., 36
Schwartz, J., 143
Scott, A., 20, 26, 28
Seidman, S., 20, 24, 26, 28
serial killers, 167–8
Sharp, R., 37
Shaw, C.R., 153
Shilling, C., 4, 7, 42, 65, 69, 70, 71, 96, 97, 103, 115, 121, 158, 165
Shipman, P., 113
Sibeon, R., 1, 5–8, 10–65, 76, 94, 96–102, 108, 114, 116–21, 123–4, 126–7, 131–6, 139, 141, 147, 153, 159–63, 165–7, 169–71
Sim, J., 160
Simon, W., 106, 108, 114, 118, 122, 151, 152, 154
Smart, C., 20
Smelser, N., 35, 41
Smith, M., 13
Social Theory and Human Biotechnology (Owen), 167
Soja, E., 120
Sparks, R., 139
Starr, H., 46
Stoker, G., 22
Stones, R., 10, 13, 27
Sutherland, E.H., 153
Swingewood, A., 26
Symons, D., 82

Tattersall, I., 113
Taylor, S., 66
Thain, M., 86
time-space
  generally, 57–61, 120, 169
  variability, 22
Tomsen, S., 155
Tooby, J., 2, 7, 12, 82, 84, 88, 89, 90, 114, 142, 146, 147, 155, 170
Touraine, A., 16
Trivers, R., 93
Turner, B.S., 13, 25

underclass, 117, 131

Van Gelder, T., 79
virtual criminology, 128, 170
The Voice and the Eye: An Analysis of Social Movements (Touraine, Duff), 16
Volavka, J., 4, 158

Wainwright, D., 72
Walby, S., 151
Walker, A., 113
Walklate, S., 140

Walsh, A., 1, 4, 7, 65, 158
White, H.C., 24, 28
Wieviorka, M., 10
Wilkie, T., 145, 146, 147, 154
Wilkinson, R.G., 72, 73
Williams, G.C., 85
Williams, S.J., 12, 69, 71, 72, 73, 95, 96, 97, 103, 104, 105, 154, 155, 165
Williams-Jones, B., 124, 125, 126

Wilson, D., 10
Wilson, J.Q., 4, 158
Wilson, M., 4, 7, 10, 12, 76, 84, 91–2, 94
Winlow, S., 10, 155

Yar, M., 10
Yeates, N., 138
Young, A., 20
Young, J., 159

Printed and bound in the United States of America